Maivé
2012

Dublin Docklands Reinvented

THE MAKING OF DUBLIN CITY

Series Editors
Joseph Brady and Anngret Simms
School of Geography, Planning & Environmental Policy
University College Dublin

Joseph Brady and Anngret Simms (eds), *Dublin through space and time, c.900–1900*

Ruth McManus, *Dublin, 1910–1940: shaping the city and suburbs*

Gary A. Boyd, *Dublin, 1745–1922: hospitals, spectacle and vice*

Niamh Moore, *Dublin docklands reinvented*

Joseph Brady, *Living in the city, 1940–2000: a social and economic geography* (forthcoming)

Dublin Docklands Reinvented

THE POST-INDUSTRIAL REGENERATION OF A EUROPEAN CITY QUARTER

Niamh Moore

FOUR COURTS PRESS

Set in 11 pt on 14 pt Garamond by
Carrigboy Typesetting Services, for
FOUR COURTS PRESS LTD
7 Malpas Street, Dublin 8, Ireland
e-mail: info@fourcourtspress.ie
http://www.fourcourtspress.ie
and in North America for
FOUR COURTS PRESS
c/o ISBS, 920 NE 58th Avenue, Suite 300, Portland, OR 97213.

© Niamh Moore and the editors 2008

ISBN 978–1–85182–834–0 hbk
ISBN 978–1–85182–835–7 pbk

All rights reserved. No part of this publication may be
reproduced, stored in or introduced into a retrieval system,
or transmitted, in any form or by any means (electronic, mechanical,
photocopying, recording or otherwise), without the prior
written permission of both the copyright owner and
publisher of this book.

Printed in England
by MPG Books, Bodmin, Cornwall.

Contents

7
LIST OF ABBREVIATIONS

9
ACKNOWLEDGMENTS

11
SERIES EDITORS' INTRODUCTION

15
THE CITY AND THE SEA

39
DUBLIN'S WATERFRONT IN GLOBAL PERSPECTIVE

The transformation of the urban economy – The impact on the urban waterfront – An international stereotype? Dublin in the early 1980s – Responding to change: the waterfront as urban playground?

66
THE POLITICS OF PLANNING DOCKLANDS, 1980–1987

The Irish planning framework – Redevelopment plans of the Dublin Port and Docks Board – The need for regeneration – The Gregory Deal – Urban Renewal Act, 1986 – The Custom House Docks: a unique planning environment – Internationalising the city: the vision for the Custom House Docks

109
RECREATING THE WATERFRONT: THE CUSTOM HOUSE DOCKS AND ENVIRONS, 1987–1997

Genesis of the International Financial Services Centre – Turning the corner: the IFSC post-1990 – Financial Services: the engine of the Celtic Tiger? – Benefits and costs of the incentive-led approach – Other elements of the Custom House Docks scheme – The changing residential structure – 'Socio-economic cleansing' at Sheriff Street – A need for change: the establishment of the Dublin Docklands Development Authority – The 1997 Master Plan for Docklands – Managing the future – A decade of redevelopment

163
CREATING A LIVING CITY: CHANGING DIRECTIONS FOR DUBLIN DOCKLANDS

A changing population and place – Promoting the 'new docklands' – Regeneration and 'community capacity' – Education in docklands – Docklands: a campus without walls – Accommodating development – A model for other cities?

213
THE CONTESTED CITY: DOCKLANDS IN THE NEW MILLENNIUM

Heritage and development – Stack A: an ingenious construction – A science centre for the city – The Museum of Dublin – The politics of development – Spencer Dock: The proposed development – Planning in a democratic state? – The politics of the Dublin waterfront – Dublin's new city quarter: Spencer Dock today – Environment and development – Remediation and redevelopment at the Grand Canal Dock – Future development: The contested nature of the Poolbeg Peninsula – Questioning development?

286
CONTINUING EVOLUTION
The challenges of redevelopment – The controlled but creative city – Future challenges

301
BIBLIOGRAPHY

304
LIST OF ILLUSTRATIONS

309
INDEX

Abbreviations

AOL	America On-line
CBD	central business district
CCTV	closed circuit television
CHDA	Custom House Docks Area
CHDDA	Custom House Docks Development Authority
CHDDC	Custom House Docks Development Company
CIÉ	Coras Iompair Éireann
CLC	Community Liaison Committee
DABS	Docklands Area Bus Service
DART	Dublin Area Rapid Transit
DDA	Dublin Docklands Area
DDDA	Dublin Docklands Development Authority
ENFO	Environment Information Service
ESRI	Economic and Social Research Institute
ETC	environmental traffic cell
FINEX	Financial Exchange
ICSH	Irish Council for Social Housing
IDA	Industrial Development Authority
IFSC	International Financial Services Centre
LDDC	London Docklands Development Corporation
NAPS	National Anti-poverty Strategy
NAPSincl	National Action Plan on Poverty and Social Inclusion
NCI	National College of Ireland
NYCE	New York Cotton Exchange
QUANGO	quasi non-governmental organisation
TD	Teachta Dála (member of parliament)

Acknowledgments

The origins of this book lie far from Dublin in a trip I took one sunny afternoon in 1996. I was working in New York and on my day off wandered to a favourite part of the city, the regenerated waterfront of South Street Seaport, to have a look around the shops and the historic ships. The maritime feel of the place, the sense of the old and the new co-existing, and the possibilities for the future that existed there really intrigued me. Coincidentally a few days later a letter arrived from Professor Anngret Simms, Acting Head of the UCD Department of Geography at the time, offering me a place on the postgraduate programme. As an aside she mentioned that she had spent an interesting Sunday afternoon at the International Financial Services Centre in Dublin looking at the new waterfront renewal scheme under construction and she suggested that it might be worth considering as a research topic on my return home. It seemed like such a coincidence that I had to explore the suggestion and I have since spent a lot of my time researching and writing on the changes in Dublin Docklands. Little did Anngret know then that she would eventually become joint editor of a book on that very part of the city.

For the initial spark of an idea and her more recent, steadfast editorial work I would like to sincerely thank Professor Anngret Simms. For his editing, visual skills and access to a map and photographic collection of Dublin that is remarkable in its scope and quality, I owe a large debt of gratitude to Dr Joe Brady. I would also like to thank most sincerely and acknowledge the continuous support of Dr Tony Parker, who not only advised me on the academic side of my work but has also been a source of guidance through the PhD thesis, postdoctoral research, as a colleague at the School of Geography, Planning and Environmental Policy and more recently as a friend. To all my colleagues, but especially to Mary Gilmartin and Gerald Mills, thank you for your encouragement and support. Special thanks must go to Stephen Hannon for employing his wonderful cartographic skills to produce many of the maps and assisting with other illustrations and to Stephanie Halpin for her support with all things IT-related.

For access to archives, files, reports and for giving up their time to talk to me and provide advice on various different aspects of this book, I would like to thank: Peadar Caffrey, Brid Ní Shé and colleagues at the Department of

Arts, Sport and Tourism; Terry Durney, Peter Coyne, Eleanor Smyth and Carmel Smyth formerly and presently at the DDDA; Ronan Kieran at the Irish Mortgage Corporation and Audrey Keegan at Hooke and MacDonald; Irene O'Gorman, Professor Joyce O'Connor and the staff of the President's Office currently and formerly at the National College of Ireland; Niall Dardis and Alison Nolan at the Dublin Port Archive, National College of Ireland; Aidan Brady, Chief Executive at Citigroup; Rose & Seana Kevany, DISCovery project; Tony Gregory, TD; Ruairi Quinn, TD and Denise Rogers; Dick Gleeson and Jim Keogan at Dublin City Council; Dr Declan Redmond and Professor Brendan Walsh, UCD School of Geography, Planning and Environmental Policy; Mairéad McGrath; Dónall Curtin; Seanie Lambe; Catherine Telton at Techniquest Cardiff; Lorna Kelly, Sheila Fanning and all of the other local residents that I spoke to on an informal basis as I went about my fieldwork in docklands. Special thanks also to Dr John Bowman. Research grants from Urban Institute Ireland and PhD and Postdoctoral Scholarships from the IRCHSS made much of the initial research possible.

Finishing this book was made bearable by the knowledge that there is life after docklands and I wish to extend a sincere thanks to all my friends but particularly to Sinéad Smith, Michael Dempsey, David Jones, Declan Fahie, Vincent Daly, Ann Marie Smith and Yvonne Whelan.

To my sister, Ciara, thanks for encouraging me even when you thought I was completely mad and for giving me the inside track on working and socialising in docklands! Special thanks go to Jonathan for his continuous support, encouragement, understanding, distractions, humour and advice throughout. And finally a heartfelt thank you to my parents who spent so many years encouraging both Ciara and me to be the best we can. For all the lifts to school, college and elsewhere, the hours spent helping with homework (and more recently home decorating), for making 'crises' seem so much more manageable, and for the knowledge that even when things seem really bad we have two unwavering supporters, this book is dedicated to you.

Series editors' introduction

This is the fourth volume in a series of books, entitled *The Making of Dublin City,* that describe the development of Dublin from the earliest times to the present-day. Our perspective is geographical. We see the city in a way that is holistic and inclusive and allows us to communicate a sense of Dublin as a living, breathing and complex entity to our readers. Central to this has been a focus on the streetscape and the forces that have shaped it, whether it be the determined efforts of the Wide Streets Commissioners in the eighteenth and nineteenth centuries or the accidental effects of the 1916 Rebellion and the Civil War. We realise that the city's landscape exists in three dimensions and that it is difficult to convey a sense of place in words alone. A hallmark of our volumes therefore has been an emphasis on visual imagery and the books are illustrated by many graphics, drawings and photographs and are visual records in their own right. In producing this series, we remain immensely appreciative of the support of Four Courts Press, especially Dr Michael Adams and Martin Fanning. They had the courage to take on the series when it was far from the established entity it is today and without them it would not have been possible.

We decided that the series would comprise books of two kinds. The first two volumes looked at the city's development in a chronological fashion and we will return to this approach in the fifth book in the series when we will examine the development of the city centre in the early to mid-twentieth century. The first volume in the series, *Dublin through Space and Time*, discussed at significant episodes in the growth of the city from its origins as a small Viking trading base to the beginnings of the twentieth century when it stood on the brink of its enormous suburban expansion. This suburban expansion was the focus of the second volume, *Dublin, 1910–1940*, in which Ruth McManus looked at the early suburban housing schemes and the complex relationships between public and private sectors. Happily both texts are still generally available and, while they stand on their own merits, also provide a solid foundation for the second group of texts. These are thematic in approach and explore some aspect of the city in more detail as for example the third volume, *Dublin, 1745–1922: Hospitals, Spectacle and Vice* by Gary Boyd. This book took as its starting point the city of Dublin in what is sometimes described as its golden age when the Gardiners, Fitzwilliams and later the Wide Streets Commissioners were creating a capital city that stood its ground against other capitals in Europe. It was a city where hospitals for the poor were supported by public entertainments frequented by the wealthy but where the displays of public virtue masked private sexual adventures. The book provided an insight into this other city.

This fourth volume by Niamh Moore, *Dublin Docklands Reinvented*, takes us into the late-twentieth century and the redevelopment of Dublin's docklands. Dublin is a port city, founded because the geography of the bay and river proved useful to Viking raiders who later became traders. However, the relationship between the port and its geography has always been difficult, a characteristic explored in our volume *Dublin through Space and Time*. While Dublin Bay with its crescent shape and broad expanse of water seems inviting, it is anything but. The bay is shallow and its sandbanks, exposed at low water, offer only a narrow channel into the river. This is further complicated by the Dublin Bar, a notorious sandbank across the opening of the river channel which while suitable for sleek shallow-draught Viking longboats caused major difficulties for trading ships sitting lower in the water. As a result, for most of its history, Dublin Port has been accessible only at high water and with the wind in the right quarter. The legacy of shipwrecks in the Bay is testament to its dangers. Likewise the river itself was broad and shallow and needed significant taming between its quay walls to make it suitable for the increasingly larger and heavier shipping. Thus the story of the port of Dublin has been a constant interaction between the changing needs of shipping and the geography of the bay. This led, *inter alia*, to the construction of the two sea walls in the eighteenth and early-nineteenth century that finally tamed the Dublin Bar and continue to keep it in check today. It also set the port on its drift eastwards in search of deeper water and more space to accommodate changes in maritime technology.

By the end of the nineteenth century, Dublin Port was beyond Butt Bridge and the city had moved eastwards in tandem with the port. However, during the twentieth century while the port continued to move eastwards, as it still does, the city did not follow. Perhaps the simplest reason for this is that the railways, and especially the Loop Line, provided both a physical and a psychological barrier. This occurred in other cities too, especially port cities, where a sense of the 'wrong side of the tracks' developed for locations beyond the sidings and goods yards and the paraphernalia of railways. Yet the port continued to change and its technology continued to evolve. It gradually abandoned land to the east of the Custom House as this became surplus to its needs. This land struggled to find alternative uses and gradually decayed into a landscape of warehouses, poor housing and half-abandoned freight yards. This was already evident in the 1930s when the port authorities advertised the ready availability of sites within the docklands for factories and warehousing. By the 1960s, there was a substantial landscape of extensive land uses to the east of the city of Dublin that was probably unknown to most Dubliners as there was little reason to bring them there. This is the point at which Niamh Moore begins her discussion of the revitalization of Dublin's Docklands. A number of forces underpin this current transformation. The growth of a global economy has provided the opportunity for cities such as Dublin to become players in industries that would not otherwise have located in such a peripheral location. The re-imaging of docklands in many other parts of the world has provided the impetus to planners and developers to see the potential of waterfront sites for culture, commerce and leisure. The existence of a large bank of underutilised land has made it all possible. As a result, a new city is developing on land that, less than twenty years ago, seemed destined to decay into dereliction. It is a city that reflects a new international identity for Dublin in the international style of its landscape. However, it is not simply a copy of what is being done elsewhere; it has its own character – that which makes it Dublin's interpretation of global phenomena. Niamh Moore explores the processes that resulted in the landscape of today. She looks at the movers and shakers and traces why the result has taken on this particular shape and not another. It is an unfinished story. The process of change and development is still continuing but the shape of the modern city is now clear for us all to see.

JOSEPH BRADY AND ANNGRET SIMMS

The city and the sea

> The history of the Dublin Docklands can be read as a microcosm of the history of Ireland, both ancient and modern. The making of the Docklands itself is a complex tapestry of great engineering achievement, visionary planning, intrigue, economic rise and decline, and human triumph over adversity.
>
> (DDDA, 1997, p. 18.)

Dublin docklands has only recently become a focus of intrigue for those with a general interest in the evolution of Ireland's capital city. Yet, it has through its history been a microcosm or laboratory in which the results of the various interactions between planning, engineering, economic change and political decision-making, themes to which this book will later turn, have been played out. Formerly the heart of the historic port and somewhat at the periphery of mainstream urban life, the present-day docklands district is now the centre of the Irish financial world and plays host to innumerable trendy bars and cafés, as well as being a hive of student activity. While it is well known by those from the world of business, for many Dubliners and others, it remains relatively unknown territory and has yet to be integrated into their activity and mental maps of the city. This book aims to contribute to this process by telling the recent story of Dublin docklands, focusing on the radical transformations in the physical and social fabric of this area in the last quarter-century. Before this, Dublin docklands was primarily a maritime district facilitating the activities of the port that has played such a central role in the development of Dublin for over one thousand years. Often perceived as not quite part of the city, the fortunes of the port have always been historically closely connected to the fate of the city, nowhere more evident than in the transitional docklands area that acted as a link between the commercial functions of the city of Dublin and the maritime functions of its port.

This book does not intend to provide a comprehensive history of Dublin port as excellent work has already been undertaken by others (Gilligan, 1988; O'Donovan, 1986), but it is useful to set the historic context for the present day development in Dublin docklands. Unlike at the present time when Dublin port is located at a significant distance from the city centre and operates almost as an entirely separate entity, Dublin in the medieval period could rightly have been termed a city-port because of the symbiotic relationship between these two realms. The general trend in Dublin, in common with other ports around the world (Hoyle, 1994), has been a separation of these entities and the result has been a progressive eastward development of the port over time creating a gap or 'interstitial area' between the city and the sea. If we had visited Dublin in the seventeenth century, we would find the heart of the city much further upstream than the present central axis focused on O'Connell Street. Close to the present-day location of the Civic Offices and the Four Courts, the city fathers and citizens developed a thriving commercial port. This was not achieved without difficulty, for Dublin is far from being a natural port: it is tidal, rocky and had a tendency to silt-up very quickly, endangering those who plied their trade along its course (see Simms in Brady and Simms, 2001). Even in the Anglo-Norman period, the river posed many difficulties and the city residents were forced into modifying the shape of the river channel. The river channel had many different pools, inlets and bays, as discussed by de Courcy (2000). Using archaeological evidence and data from engineering cores, he offers a picture of a river with its slow S-bend established early on in the history of the city. There were bays and pools near Christ Church on both sides of the river, one of which had to be filled in by Ellis in his seventeenth-century reclamation, which is mentioned below. Beyond what is now O'Connell Bridge, the river widened dramatically as shown in Figure 1. The broad sweep of the bay belied the fact that it held numerous dangerous sandbanks that made the approach to the city hazardous. Added to this was the infamous Dublin Bar, a sandbank that effectively cut off the city from the sea at low tide and which ultimately required the great engineering works of the North and South walls to tame. These sandbanks can be seen clearly in Figure 2 which is taken from Greenville Collins' *Great Britain Coasting Pilot*, first published in 1693. The geography of both bay and channel was a trial to the merchants of the city from early times. We have the complaint of one merchant to Edward III in 1358 that 'from want of deep water in the harbour … there has never been anchorage for large ships from abroad' (quoted in Gilligan, 1988, p. 9).

Managing and taming the Liffey has been an enterprize that is as old as the city and we should not be surprised that it continues to the present day. An examination of early maps of the city shows that extensive land reclamation occurred on both sides of the river channel and the formalization of quays along the northern edge of the town wall in an attempt to improve navigation

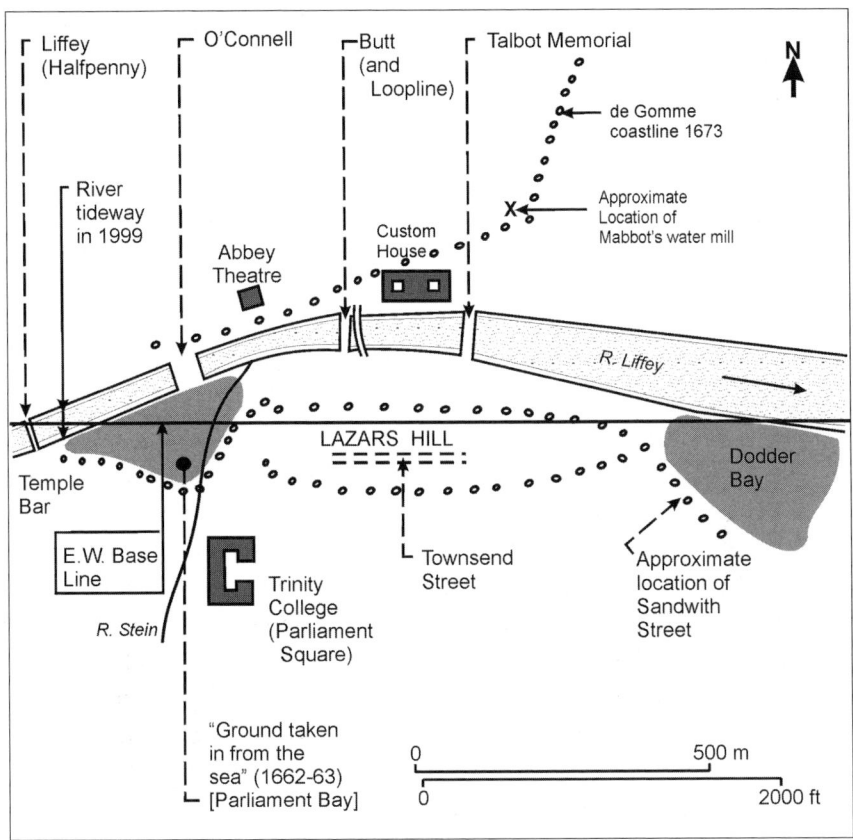

1 The Liffey channel in the seventeenth century. (De Courcy, 2000.)

and berthage. The problems caused by the absence of a natural harbour continued and by the late-seventeenth century the idea of altering the river channel permanently became a key enterprize. Artificial quays including 'Wood Key', 'Blind Key' and 'Custom House Key' were constructed on the south side and later Ellis Quay was developed on the north side of the river (Figure 3). Many of these quays remain in the contemporary city and the buildings that developed along their length have been the subject of extensive urban renewal in recent decades.

These infrastructural developments probably facilitated some increase in the volume of activity through the port resulting in the widening of Blind Quay (east of Wood Quay) in 1684, yet by 1672 problems remained and the river was still only navigable at spring tides (Gilligan, 1988). One attempt to circumvent this problem was the gradual eastward re-location of the port into

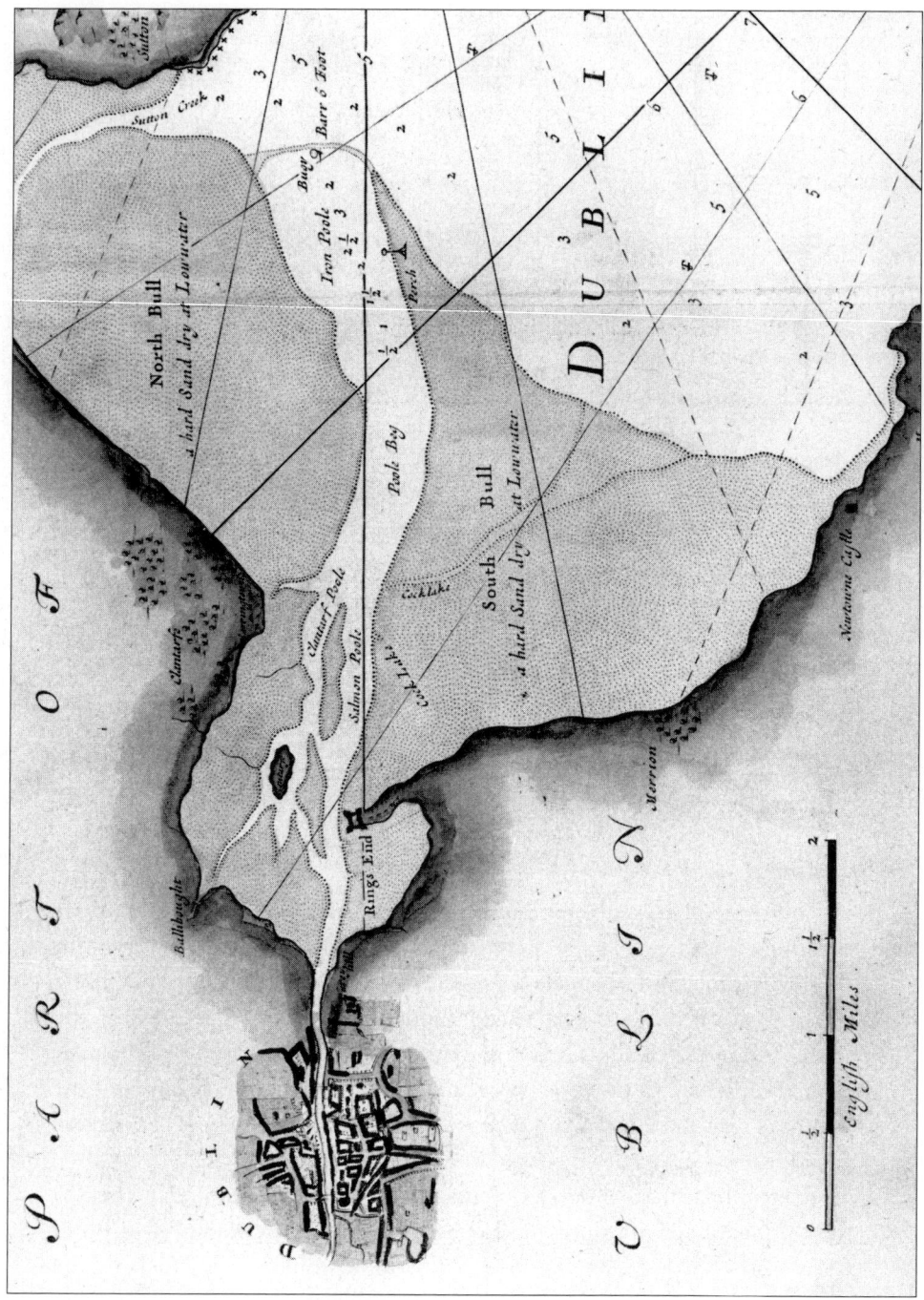

2 The bay and harbour of Dublin.
(Surveyed by Captain Greenville Collins, 1686.)

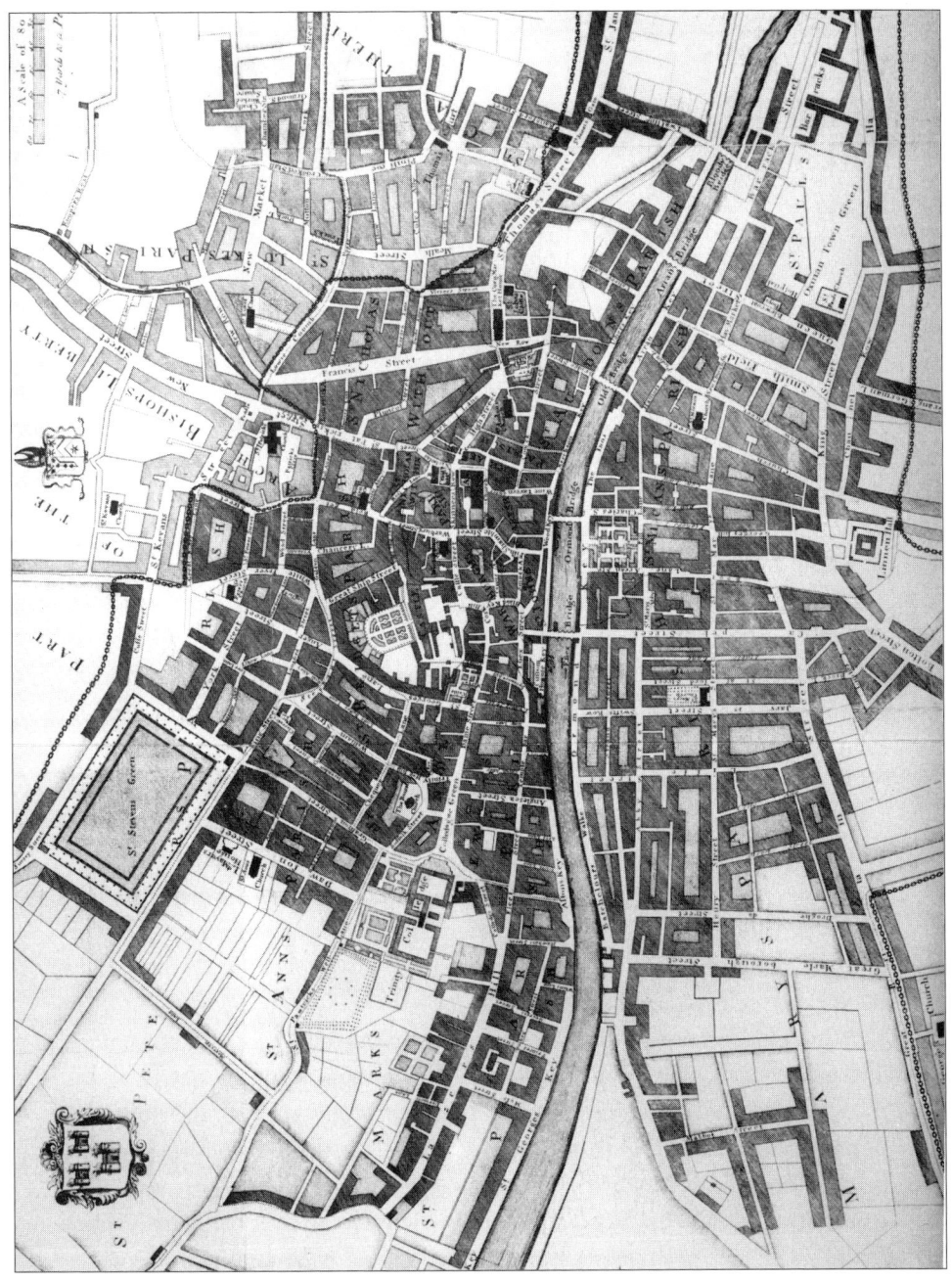

3 The development of Dublin's quays by 1728. (Brooking, 1728.)
(Note the orientation of the original map is south to north.)

4 The Queen Anne Custom House. (Brooking, 1728.)

deeper water, evidence of which is provided by the changing location of the Custom House, the focal point of all port activity. The first Custom House was built in 1620 and replaced in 1637, and a third further downstream was completed in the 1660s. The predecessor of the present Custom House was constructed in 1707 and located on the site of the Clarence Hotel at Wellington Quay, although the quay as we know it was not completed until 1817. This is evident in maps of the city, including that drawn by Brooking in 1729 which illustrates a break in the continuity of the quays between the old 'Custom House Key' and 'Astons Key' (Figure 3). The Old Custom House was built in Queen Anne style, with a ground floor open arcade where business could be done (Figure 4). This is similar to the arcades still evident in other European towns and cities today, such as Bologna (Italy), Prague (Czech Republic) and Treviso (Italy). The vignette also shows the nature of the housing along the quays and it is clear that the city had yet to embrace fully the idea of focusing on the river.

While this imposing structure marked the pioneering eastward movement of the port to facilitate trade, it was still not enough to allay concerns about the economic future of the port. A number of proposals were developed to improve the situation. One of the most radical was the idea that trade could be developed and grown if the river was contained within two walls. By constraining so much water into a narrow channel, the river bed would scour itself and therefore the navigation channel would become much deeper. A 'Map of the Harbour of Dublin from Essex Bridge to the Barr, 1704' in the Dublin Port Company archive pre-dates the establishment of an authority to oversee the development of the port, but illustrates the kind of proposals

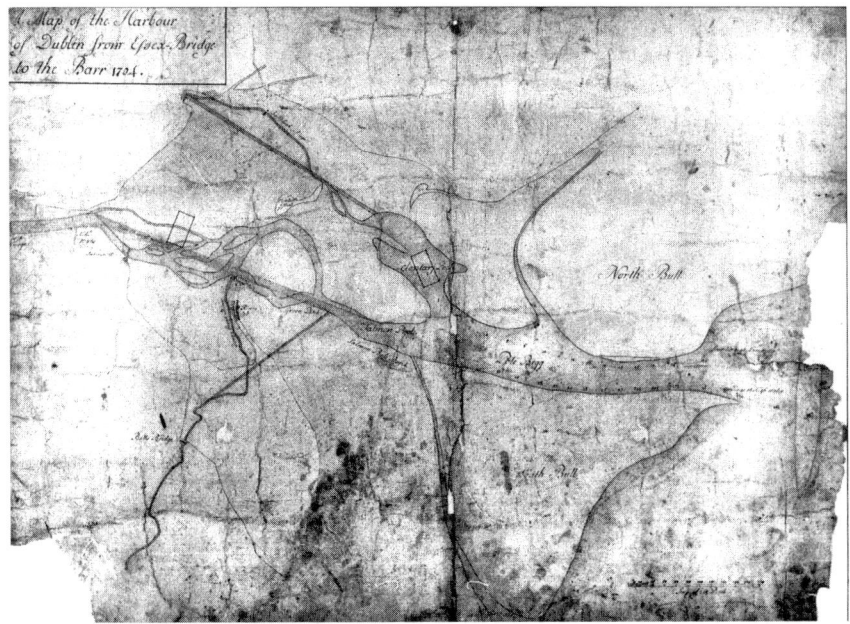

5 Map of the Harbour of Dublin from Essex Bridge to the Barr, 1704. (Dublin Port Company Archive.)

forwarded by interested parties in a bid to solve the difficulties and protect the economic viability of the city (Figure 5). It depicts the proposed line of a new channel that could be created by straightening the river Liffey as far as Ringsend, with the overall purpose of removing the natural braided channel, which was a frequent cause of shipwrecks and thus a threat to economic prosperity and urban development (de Courcy, 2000). Though it is impossible to tell, it is estimated that over 200 ships were wrecked over the centuries on the approach to Dublin port, and the route of the Dublin Bay sewage pipeline, due for completion in June 2008, had to be altered to take account of this debris in the Bay.

In 1716 work began on the construction of a wall on the south side of the channel from Ringsend to Poolbeg, clearly indicated in Figure 6. On the north side of the river, the North Bull Wall was constructed. Rocque's map of Dublin in 1756 illustrates that by that time the East Wall had been at least partially completed, and the location at which the North and East Wall met became known as the Point. This is the present-day location of the popular entertainment venue owned by Harry Crosbie, the Point Depot, which is currently at the centre of significant proposals for redevelopment as the Point Village.

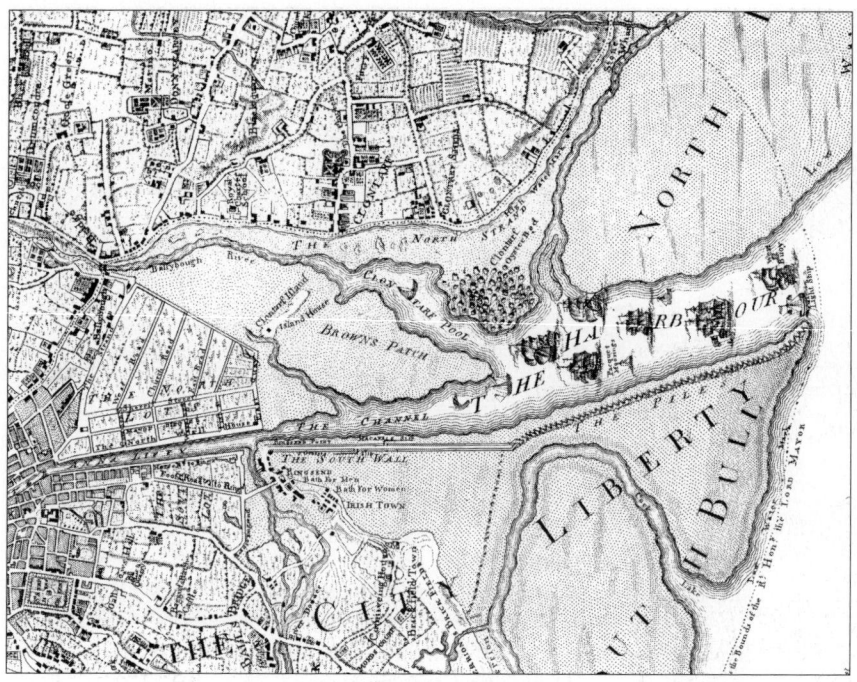

6 The beginnings of major engineering works on the river channel. (Rocque, 1756.)

While improving safety within the port was of paramount importance, these infrastructural changes also provided other opportunities for general urban development. The construction of these walls led to the reclamation and creation of new land east of the pre-existing urbanized area. These slob lands, captured between the north and east walls, had been surveyed in 1717 and divided into allotments or Lotts by the city council – giving rise to the name, North Lotts, that survives to this day. New streets were laid out in a grid-like pattern at regular intervals, and street names honoured different agents of power in the city council at the time: the mayor, sheriff, guilds and the commons (Cosgrove, 1909). The area is shown on the Ordnance Survey plan of 1836, and it illustrates the relatively slow pace of development. There is little development to the east of the Grand Canal Docks save the Vitriol Works on Mayor Street (Figure 7). This district was often referred to as Newfoundland as it had been newly created from the sea, and until the early 1900s a Newfoundland Street ran parallel to Sheriff Street and Mayor Street (Figure 8).

This migration of the port eastward transformed the relationship between city and port on the northern side of the river as well as on the south side. The

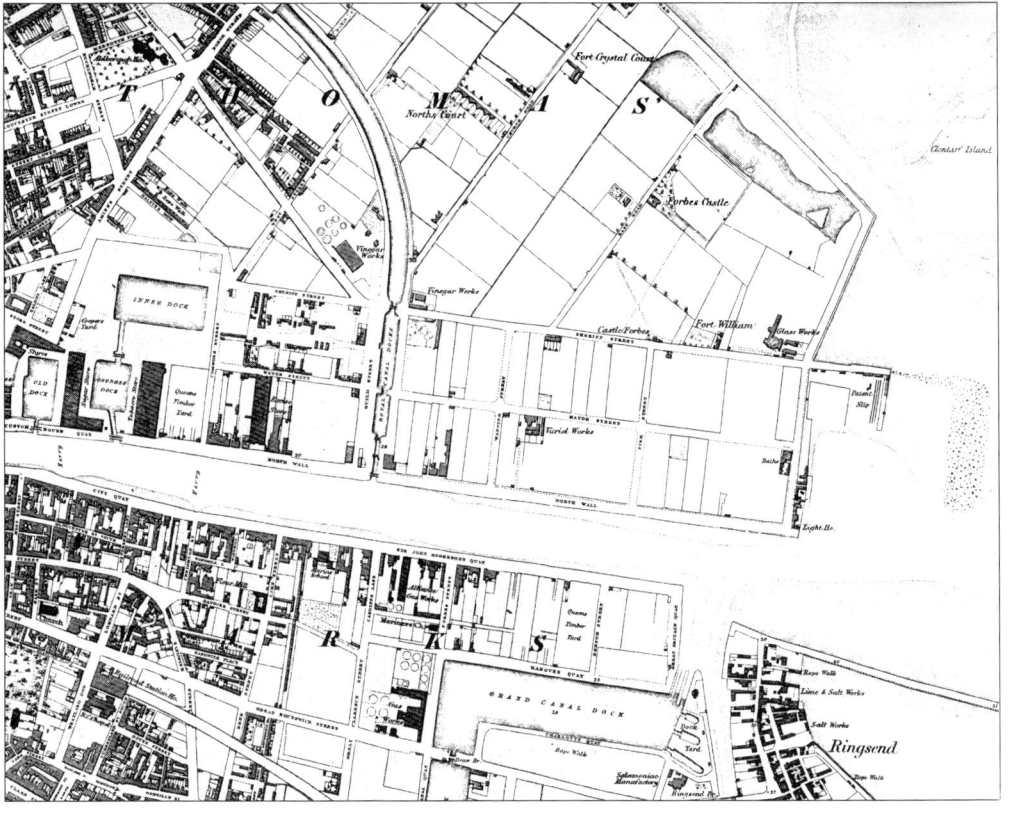

7 The 'lotts' and the docklands in 1836. (Ordnance Survey 1:10,560 plan, Sheet 18, Dublin, 1837/8.)

construction of Sir John Rogerson's Quay in 1728 provided a clear indication that the focus of commercial activity was beginning to shift seaward away from the medieval city and on to newly reclaimed 'ground taken in from the Sea' as described on de Gomme's map of the city (1673). In the early-eighteenth century houses were constructed along the length of Sir John Rogerson's Quay, even while water remained behind it. One of the Malton prints of Dublin looking up the Liffey illustrates the Marine school at Sir John Rogerson's Quay in 1796. A view of the approach to the city with the Custom House in the far background, it shows a wide river with busy shipping and that new functions, emerging in this area, were closely related to the operation of the port. Like the area north of the river, the new land created on the south was divided into lotts to become known as the South Lotts and this is still evident in the contemporary landscape. South Lotts Road, one of the primary

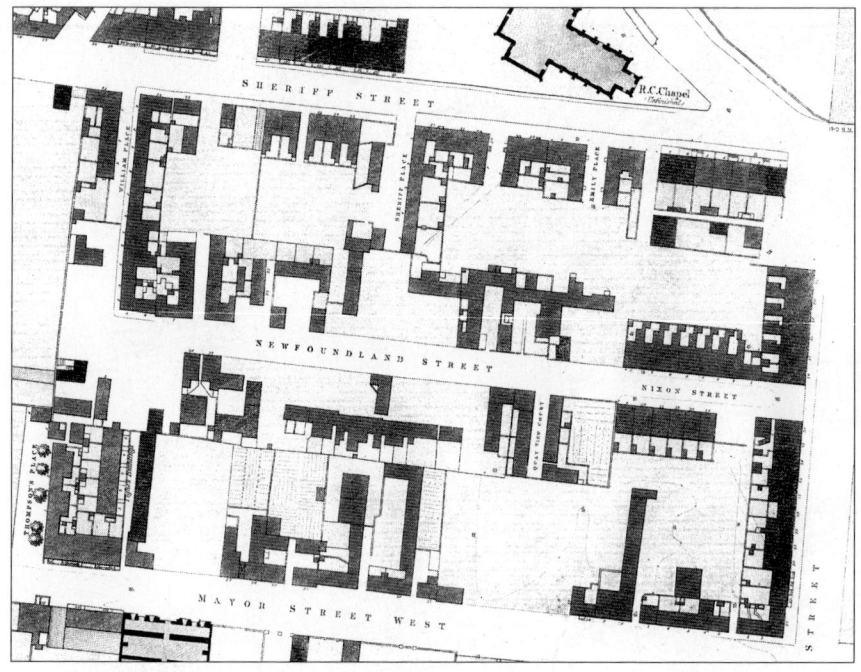

8 Newfoundland Street, 1876. (Ordnance Survey 5-foot plan, Sheet 15, City of Dublin, 1846/7.)

roadways through the former Sir John Rogerson's Strand, is an important present-day link between Ballsbridge and Ringsend-Irishtown.

The migration of the port not only provided room for expansion: other positive benefits emerged for older districts of the city. Prior to the reclamation, the village of Ringsend was frequently cut off from the city and, even after reclamation, flooding still remained a threat although a much reduced one. The following extract from the *Dublin Chronicle* (28 January 1792) illustrates how unpredictable sailing in the Bay and the river channel could be in eighteenth-century Dublin:

> 'His Grace, the Duke of Leinster, went on a sea-party, and after shooting the breach in the south wall, sailed over the low ground in the south lots and landed safely at Merrion Square'.

One of the most impressive infrastructural developments to emerge from the reclamation of land behind Sir John Rogerson's Quay was the Grand Canal Docks complex. These large-scale docks, covering 35 acres (*c.*14 ha), were opened by the earl of Camden in 1796 and represented the first purpose-

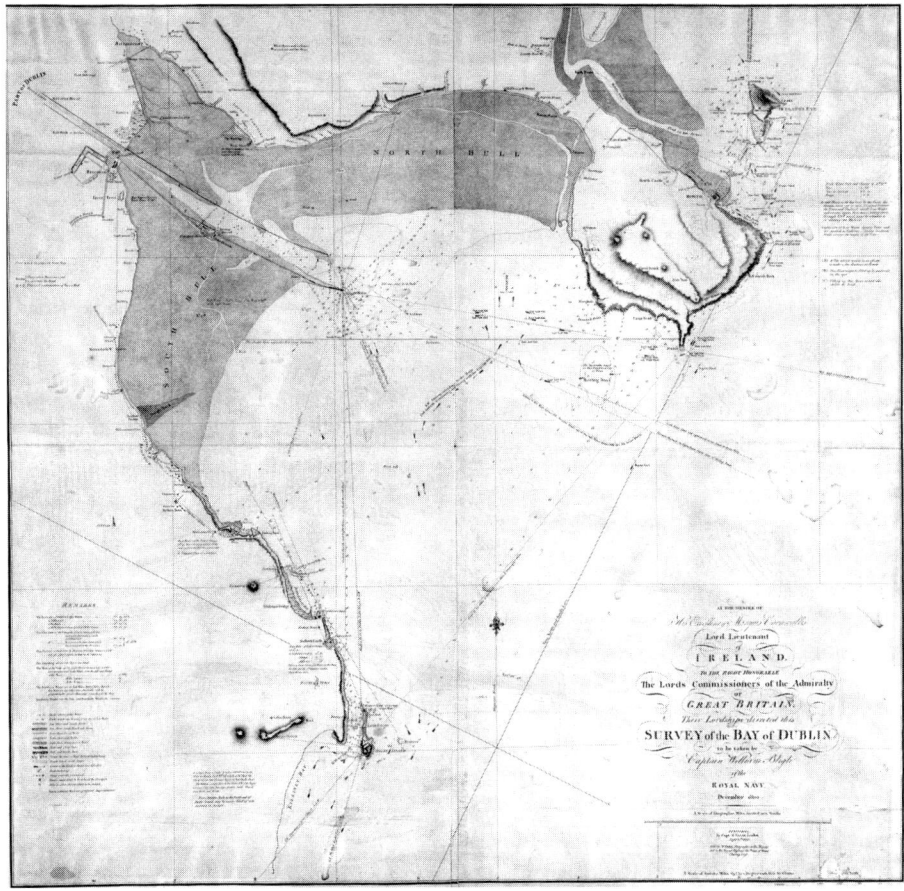

9 Map of Dublin port by Captain Bligh, 1803. (Dublin Port Company Archive.)

built docking facilities for sea-going vessels. Two large basins, constructed at an estimated cost of £112,752, are joined by a lock and open to the Liffey at the same point as the river Dodder (Maxwell, 1997). This new complex provided opportunities for growing trade but also created significant challenges as urban activity began to move away from the former core of the city to reclaimed land, a pattern that continues today. This was further compounded and intensified by the many technological and engineering achievements of the nineteenth century.

In 1800, Captain John Bligh, commander of the infamous HMS *Bounty*, was sent to Dublin by the Admiralty to undertake a survey of, and suggest improvements to, the dangerous and still problematic river channel. He

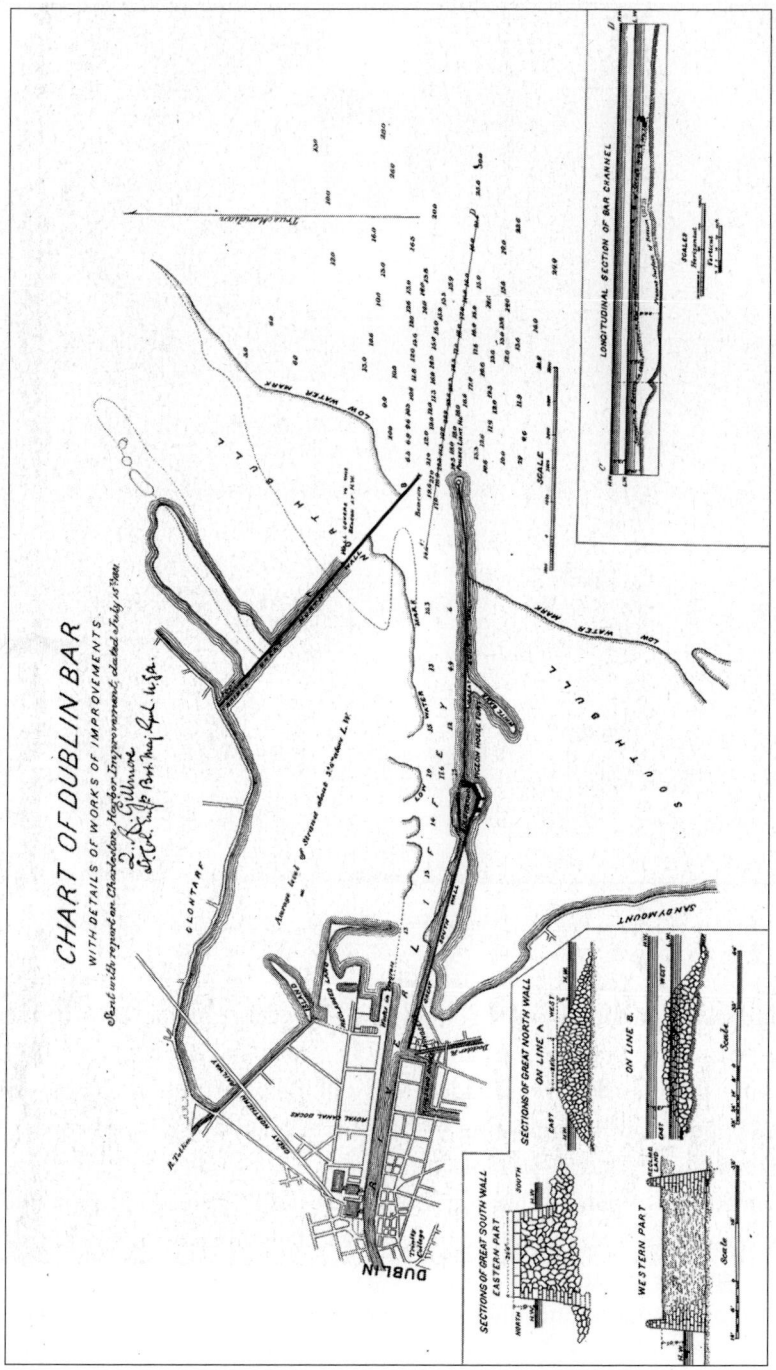

10 Chart of Dublin Bay with cross-sections of the wall, 1881.

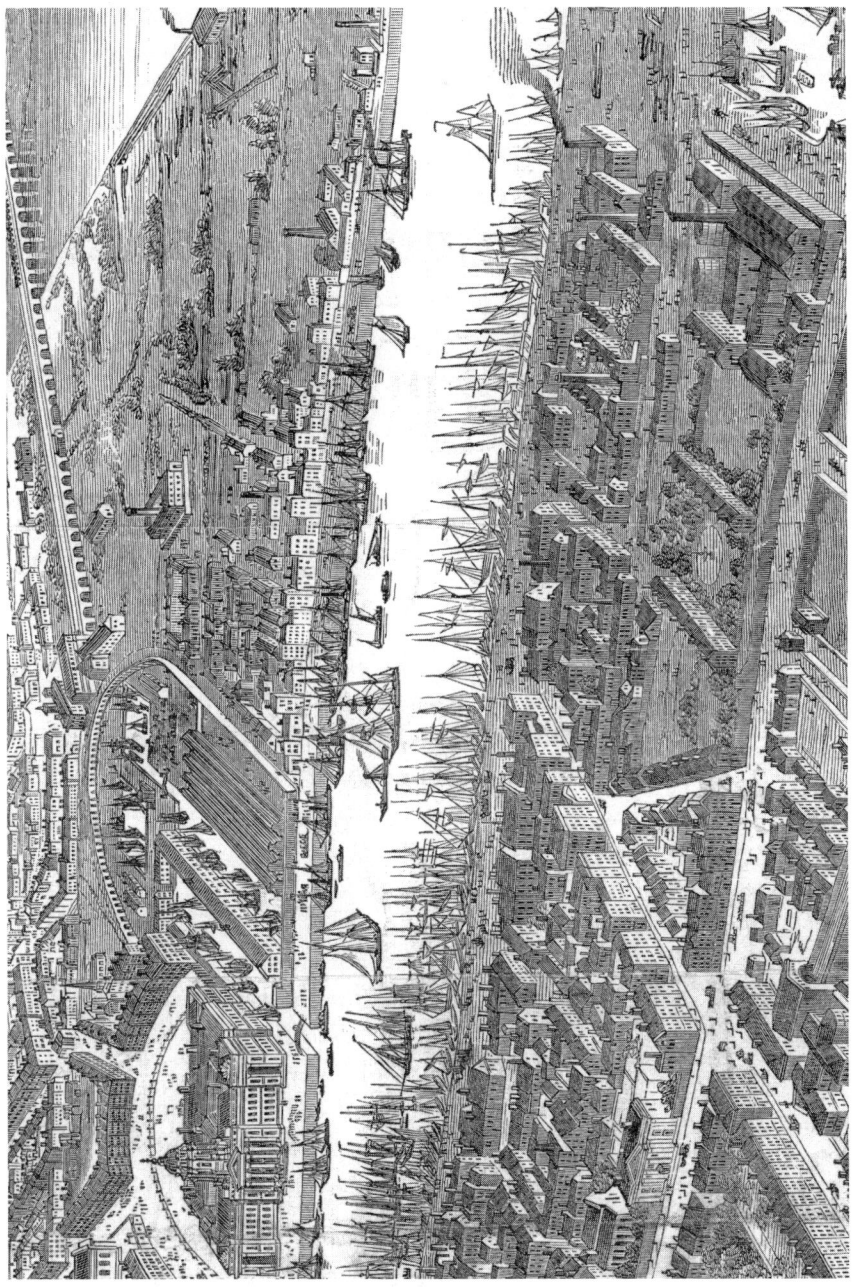

11 Dublin port in 1846. (*Illustrated London News*, supplement, 1846.)

12 Shipping near the Custom House. (Thomas Kelly, London.)

proposed the construction of a wall on the north side of the river parallel to the Great South Wall to improve the natural scouring effect of the channel (Figure 9). The wall that he proposed was never built, but it seems to have significantly influenced the thinking of Francis Giles, who came to Dublin almost twenty years later and designed the North Bull Wall. Following the completion of this major project in the 1830s, the river channel became much deeper and facilitated the docking of much larger vessels. Yet just as one issue was resolved, another began to occupy the minds of port engineers. In the late-eighteenth century, trade at Dublin port had increased dramatically requiring the further development of new dock areas for cargo loading and unloading. King George IV opened the eight-acre George's Dock in 1821 and a short time later, the much-larger Inner (or Revenue) Dock was completed. These docks have in recent years become the focal point around which the International Financial Services Centre and other commercial buildings have been constructed, a clear example of the evolutionary nature of cities and their response to changing economic demands. Unfortunately from the time they were constructed these docks were virtually obsolete as rapid changes in shipping technology led to the growth of larger vessels that simply could not fit through the narrow entrance into these sheltered water bodies. This failure of planning to keep pace with technological developments might be compared with the current situation in Dublin where the new port tunnel is simply not large enough to cope with large super-trucks that are accommodated quite easily on the European mainland.

Two images from the mid-nineteenth century suggest a busy and prosperous port. The panorama produced by the *Illustrated London News* in 1846 shows quays that are filled with shipping right to the eastern edge of the port. The Inner, Custom House and George's docks are hives of activity and there is congestion on the George's and City quaysides. The engraving by Thomas Kelly, published in London, focuses on the area around the Custom House and shows a variety of ships, big and small, in a busy scene that includes both sail and steam ships.

However, care must be taken with this impression. The text accompanying the *Illustrated London News* makes for less positive reading. It suggests that the Custom House, fine edifice that it is, might be considered 'the cenotaph of Ireland's commerce' for the following reasons.

> When the Custom House was erected, the trade of Dublin was rapidly increasing every day and its dimensions were formed more with a view to the prospective requirements of what might be fairly looked to – judging from its then steadily and rapidly progressing commerce – as the future trade of Dublin than with reference to the actual wants of its existing commerce. But the Legislative Union interfered and stayed all further progress of the trade of the port of Dublin except in the trifling degree which is caused by the wants of an increasing impoverished population.
>
> (*Illustrated London News,* Supplement, 1846, p. xi.)

The *Graphic* magazine in its supplement on Dublin in 1878 was more upbeat. It noted that over £900,000 was collected in customs dues annually and that the prospects were bright since over £100,000 was at that moment being invested in warehouses and hotels in Holyhead in anticipation of increased trade with Dublin. Nonetheless, the Custom House itself was still in the doldrums when *Ireland in Pictures* was published in 1898. It noted that: 'since the fatal "union", only a portion of it is used for customs business'.

Despite the circumstances of trade, the port had to continue to respond to the changing demands of shipping. Like today, the fortunes of Dublin city and port in the nineteenth century were closely tied to and dependent on external trade. The rapid expansion in global trade that occurred in the second half of the century, facilitated by new technologies, placed significant demand

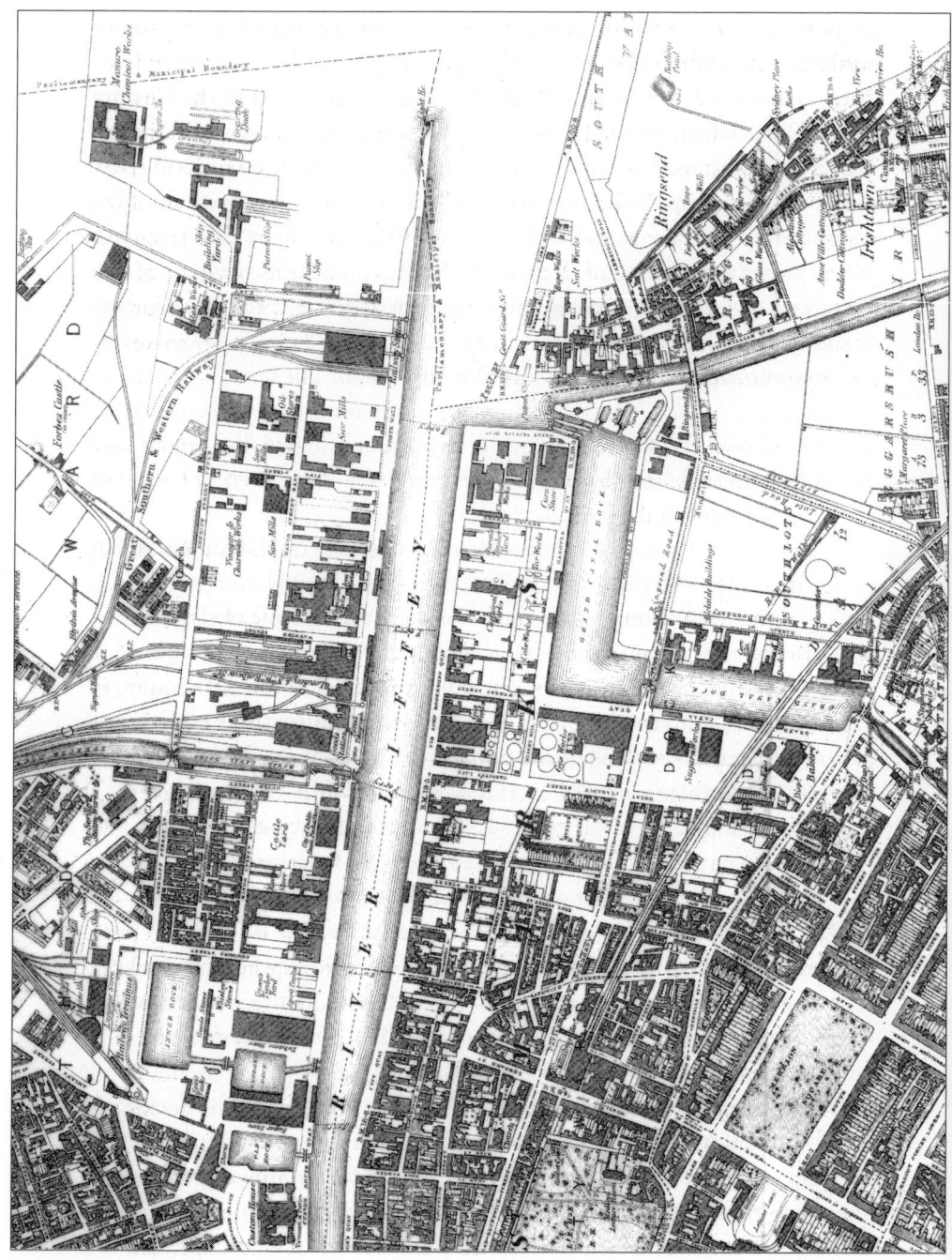

13 Port and harbour of Dublin, 1887. (Thom's Directory, Ordnance Survey 1:10,560 plan.)

on port cities like Dublin, resulting in increased demand for deep-water berthage. The Dublin Port and Docks Board responded by developing deep-water quays downstream, entirely abandoning the area around the Custom House Docks which reverted to warehousing rather than ship-based activities. This was the first indication that parts of the port, formerly considered state-of-the-art, could rapidly become obsolete and marginalized as a result of the continued demand for deep-water activities. While the North Wall extension providing 1.5 kilometres of deep-water berthage and the opening of the Alexandra Basin in 1885 had negative repercussions on port areas further upstream, these improvements facilitated the accommodation of larger vessels and repositioned Dublin port in global terms.

> The North Quay extension afforded ample room for vessels and provided tidal-berthage accommodation of a depth then unknown in most harbours throughout the world.
> (Cox, 1990, p. 21.)

This was not just a feature of development on the north side of the river. Sir John Rogerson's Quay, which had been the first speculative quayside development in the eighteenth century, continued to lead the way in pioneering new facilities and by the end of the next century had 1.2 kilometres of deep berthage, twice that of the purpose-built Alexandra Basin. This new construction enabled the port to cope with larger ship sizes and aided the economic development of the city, but it reinforced the separation of city and port that had begun in the eighteenth century. This was compounded by the construction of a port boundary wall and gates on East Wall Road that marked the physical as well as the functional separation of the city and port. However, on the positive side, large tracts of land suddenly became available for other functions and the industrialization of cities that had already taken root across Europe, found a ready home in the Dublin docklands albeit on a much smaller scale than in other places.

Entrepreneurs took advantage of the strategic location of the docklands and, by the beginning of the twentieth century, the area was permeated with industry. Coke works, chemical works, slaughter houses and gasworks occupied key locations (Figure 14). The gasworks in particular were attracted to the port because of their requirements for imported coal and coke and the need to use large quantities of water as a raw material to produce town gas. The largest complex of gasometers, developed to contain the gas, was located

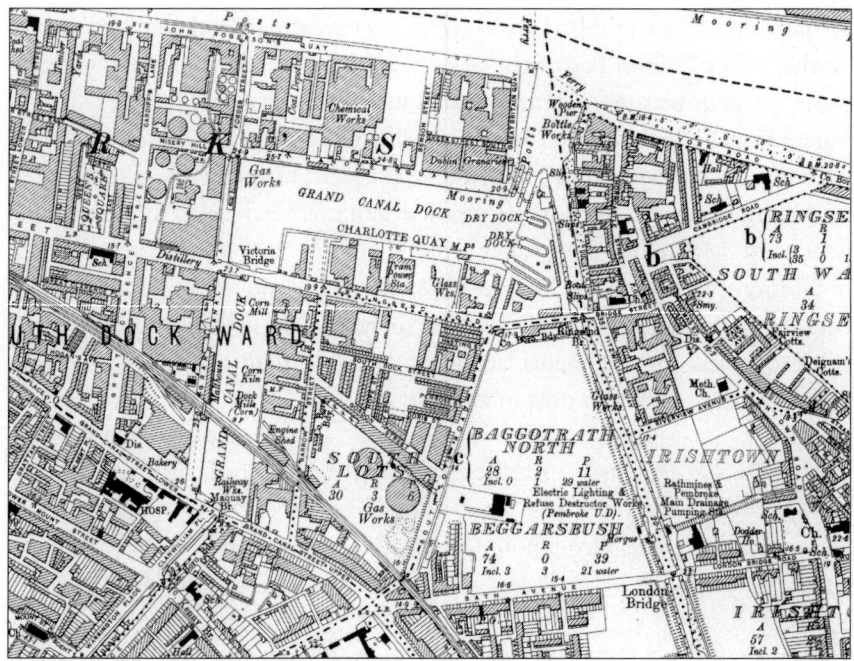

14 Industrial uses in south Docklands, 1911. (Ordnance Survey 1:10,560 plan, Sheet 18, Dublin, 1911 revision.)

between Barrow Street and South Lotts Road after land was acquired by compulsory purchase order following the passage of the Dublin Gas Act, 1866. The aesthetic value of the gasometers varied. The Clayton gasometer at Barrow Street was completed in 1871 and is visually the most impressive with its decorative wrought ironwork that has recently undergone conservation (Figure 15). It stood in stark contrast to the unattractive MAN gasometer, with a volume of 3 million cubic feet (c. 85,000 m³), that was later built at Sir John Rogerson's Quay. The production of town gas gave rise to a number of by-products that were used in developing the chemical industry, adding to the range of incompatible land uses that already existed in docklands.

By the late 1920s, the port of Dublin was handling of the order of 2.3 million tons of imported goods but less than 460,000 tons of exports, in which live stock was a crucial component. The Dublin Port and Docks Board was anxious to attract business to the port as Figure 17 below shows. Attention was drawn to the impressive facilities of the port, especially the Alexandra Basin with its 25 feet (7.6m) of water at low tide. There was ample warehouse accommodation and sites for factories with quay frontage and railway

15 Clayton gasometer immediately prior to its redevelopment, 2000. (J. Brady.)

connections. In a paper on the port in *A Book of Dublin* (1930, p. 45) it was noted that the policy of the Dublin Port and Docks Board was one of gradually entending the north quays into the bay and thus making available a large area of reclaimed land. The eastwards expansion of the port might have been necessary but it created a difficulty, obliquely referred to in the advertisement. It left a great deal of land between the city centre and the port for which it was difficult to find any uses. It was an issue in 1930 and remained so until very recently.

While economically important, the unattractive nature of this kind of industry marginalized the docklands and reinforced the separation between the city and the bay, a process that has continued into the twentieth century. The pattern of port expansion in an easterly direction in response to the need for larger tracts of land to facilitate mechanization as well as the need for deeper water to berth much larger ships, has made large areas that were once

16 Port activity in early 1920s. (Postcard.)

central to the city and port completely obsolete. Migration of the port gathered pace throughout the twentieth century and the reclamation of large tracts of land north and south of the river entirely altered the shape of Dublin Bay (Figure 18). While port facilities have not been provided in close proximity to the Custom House for many decades, the public sector did not totally abandon the area. In 1941, the State transport company (CIÉ) increased their holdings in the area by acquiring land from the Dublin Port and Docks Board to construct the central bus station, Busáras, close to the original Custom House Dock (Figure 19).

The success of this development was followed by the acquisition of land by the Department of Posts and Telegraphs in 1950 to erect a new postal sorting office at Sherriff Street, and Dublin Corporation also acquired land over the original Custom House Dock, which had been infilled in 1927, to construct Memorial Road (Figure 20).

Even in the early-twentieth century therefore, we see clear examples of public sector intervention in an attempt to bring a new role to the area that had been abandoned by the port. In a less than buoyant economic climate, these challenges proved difficult to overcome yet this transformation of the area and the attempt by various stakeholders to intervene in the development process is not unique to Dublin in the 1950s. There are many parallels today

THE PORT OF DUBLIN

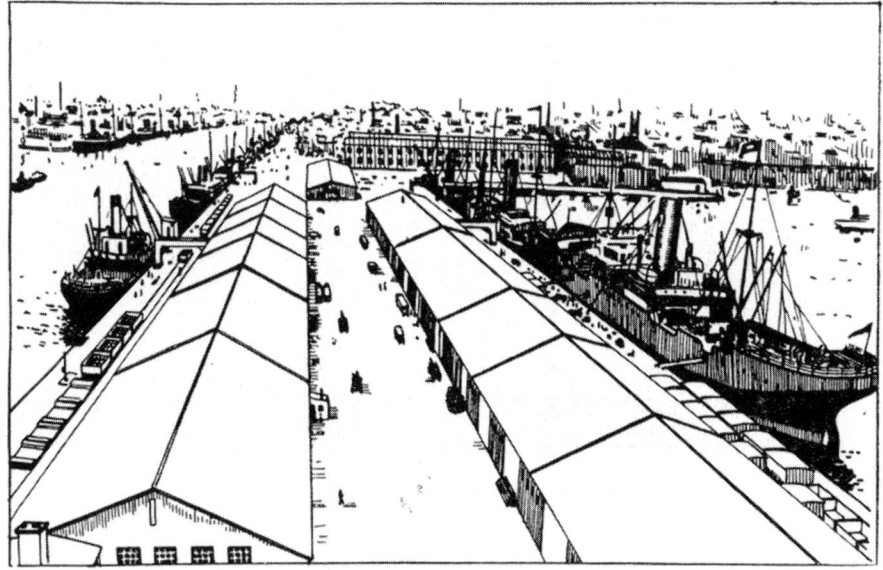

Transit Sheds on North Quays, showing Railway Sidings

Ship via the Port of Dublin
Ireland's Natural Distributing Centre

Modern Facilities for Rapid Discharge and Loading.   ::   Deepwater Berthage
Direct Communication with Railways
Sites for Factories available for letting with Deepwater Frontage and Railway Connections
Through Railway Rates.   Ample Warehouse Accommodation.   Cold Storage

Full particulars can be had from The Secretary, Dublin Port and Docks Board, 19 Westmoreland St., Dublin

17 Advertisement for the port of Dublin. (Hobson, 1930.)

with the kinds of developments that have previously taken place in the old port area, but now the private sector, as in so many other major cities, has a key role to play. In cities everywhere, local authorities are trying to find new ways to respond to changes wrought by a transformed trading environment. Increasingly they are engaging with private sector investors to devize local responses to the impacts of international economic change and decisions made by institutions and corporations at a global scale. The clear links between events in Dublin port and docklands and transformations in the global economy have continued, and some would argue have become even more intensified. Through the creation of new places of commerce and

18 Aerial view of Dublin docklands, 2004. (J. Brady.)

19 Busárus, Beresford Place, 1960s. (Postcard.)

recreation, redeveloped waterfront areas are now providing opportunities for cities to make their mark on the global stage.

While maritime activity once provided an important link and defined the relative position of cities in relation to other places, today it is the new industries, particularly the financial and cultural industries, that are colonizing

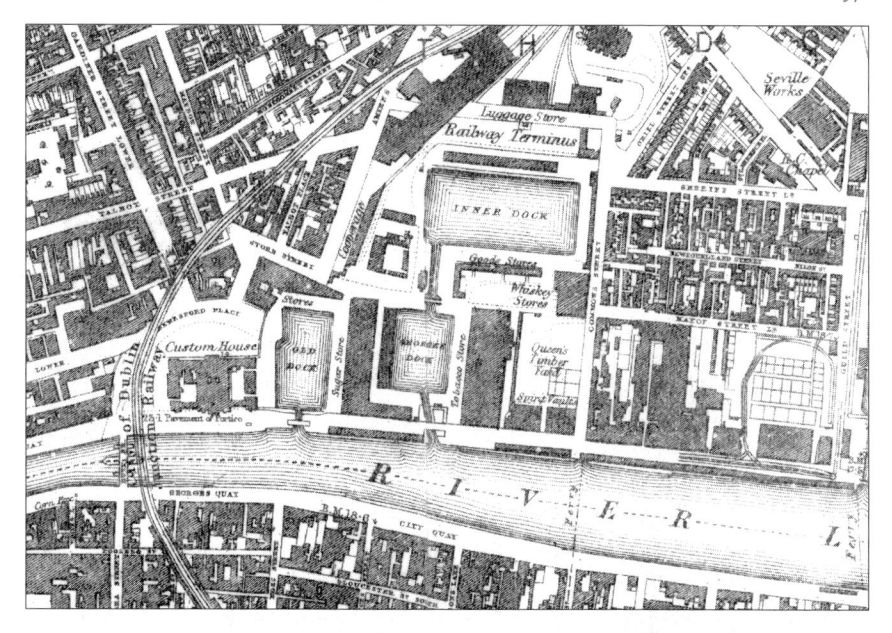

20 Old Custom House Dock (Ordnance Survey 1:10,560 plan, Sheet 18, Dublin, 1911 revision.)

these areas and facilitating global networking and linkages. In that context, Dublin docklands is not unusual, and the way in which redevelopment has been managed here bears close resemblance to similar developments in cities from Cape Town to St Petersburg, and Boston to Tokyo. Dublin docklands today is perhaps one of the most dynamic quarters in the city and is again becoming a key location through which Dublin is benefiting and perhaps helping shape the development of global trading links. Located on both sides of the river, the official extent of the Dublin Docklands Area is 1300 acres (526 ha), comprising 10 per cent of the inner city or the area between the Royal and Grand Canals (Figure 21).

While it is one of the most dynamic areas within the urban core given the scale of the opportunities it provides, it is also problematic, due to the difficult nature of development in some districts. Many commentators and stakeholders are referring to it as the panacea for all the environmental and urban ills that have resulted from the Celtic Tiger boom, a location that has the capacity simultaneously to solve the housing crisis within the Greater Dublin Region, maintain its position as engine of the Celtic Tiger economy, and provide a testing ground for policies aimed at addressing issues of marginalization and social integration within the city.

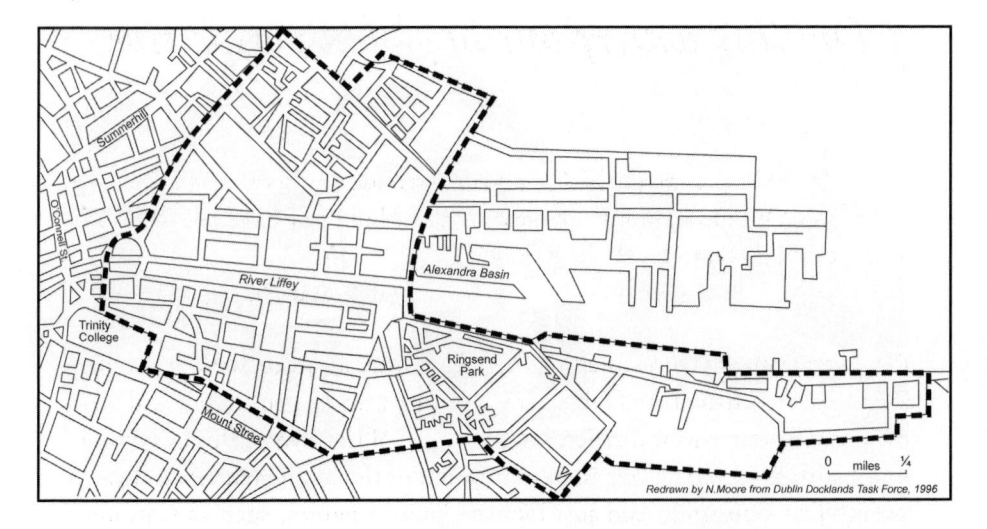

21 The extent of the Dublin Docklands Area. (N. Moore.)

The redevelopment of Dublin docklands has been continuing now for almost twenty years and it has been fraught with conflicting and changing contexts, needs, approaches and visions. The complexity of the institutional landscape within this part of the city has provided particular challenges with various disagreements arising between the Dublin Docklands Development Authority, the City Council, the Dublin Port Company, business interests and local community groups over the years. The most recent vision for the wider area, arising from a political document developed by the Progressive Democrat party, is to remove Dublin port entirely from the city and transfer activity to Bremore, close to Drogheda. The rationale is that this would open up the current port area to further commercial, residential and other more 'urban' uses continuing a trend that has been apparent now for many decades. Whether or not this should or will be undertaken is beyond the scope of this particular book. What is relevant is that the lessons of the last twenty years show how difficult it can be to promote development and balance the needs and desires of competing stakeholders in an area that is ripe for development, a difficulty not just apparent in Dublin but in most major cities that have undergone economic, social and physical restructuring.

Dublin's waterfront in global perspective

> 'In the beginning the harbour made the trade, but soon the trade began to make the harbour' (Sargent, 1938). He might have added that it could go on to break the harbour.
>
> (Hoyle and Hilling, 1984, p. 7.)

Just like the great Georgian estates that Dubliners associate with the golden age of eighteenth-century Dublin, it is likely that future generations will view the growing waterfront districts on both sides of the river Liffey as synonymous with the Celtic Tiger economic boom of the late-twentieth and early twenty-first centuries. And just like the famous names, such as Gardiner, Mountjoy, Aungier and Pembroke associated with the Georgian squares and terraces, the present-day waterfront will probably be understood as a landscape connected with the names of Haughey, Gregory, Desmond and a range of other individuals and institutions. If we want to understand fully our capital city, and indeed cities in general, then we must come to the realization that cities do not just emerge, they are not just the outcome of a range of unconnected events, but rather they are a social product (Pahl, 1975). While cities actively participate in shaping human behaviour, they also clearly reflect the societies that have shaped them. We might therefore suggest that cities act like a mirror, allowing us an insight into the kinds of individuals, societies and events that shaped them.

This is an approach that has proved effective in trying to understand change. Regeneration projects are so intriguing because they have had 'to accommodate personal political ambitions … house new nodes in the global economy' and act as 'a place where the forces of capitalism are currently exercized under a new guise' (Malone, 1996, pp 2–3). In other words we might go as far as to argue that these districts operate as a locus through which the impact of political and economic forces at a global, national and local scale are negotiated and accommodated. There is no doubt that developments in Dublin have mirrored this general global trend and it is certain that the docklands project here has become synonymous with particular local and national personalities, while simultaneously facilitating the emergence of an everyday landscape of global economic institutions in the form of international banking, finance and insurance industries. From a functional perspective, the area has become a textbook example of a district where the service or post-industrial economy has been nurtured and grown in direct contrast to the traditional activities more readily associated with port areas.

The transformation of the urban economy

A key question that then emerges is why and how this transformation has taken place? From their beginnings, cities have performed an array of functions as central places, most notably as sites of power, religion, economic development, innovation and creativity. In fact, it is exactly this diversity that allows us to characterize certain places as more successful cities than others. In the eighteenth century, cities like Dublin were important for the key decision-making or governmental role that they performed, a role that was entirely dependent on a much broader political context, in this case Dublin's role as a capital city in the British empire. The city and many of its citizens thrived in this golden age, when the important governmental role of the city spawned a range of other activities such as the demand for high quality residences, pleasure grounds and other recreational pursuits (see Sheridan in Brady and Simms, 2001). The political function of the city had a tremendous impact on circulation patterns too within the urban core, as many urban projects such as the construction of Essex Bridge and the widening of Dame Street were specifically designed to facilitate links between the residences of the parliamentarians and the Parliament Building itself on College Green. The Wide Streets Commission combined the solution to traffic problems with the creation of a capital city worthy of any monarch in the Europe of the times. After the Act of Union when Dublin's role as a capital was removed and many of the elites abandoned the city for London or elsewhere, the city experienced significant decline and decay, becoming what Daly (1984) has described as a 'deposed capital'. In a similar manner, economic changes have had profound impacts on cities and particularly those wrought by the drive towards industrialization in the nineteenth century.

At this time, the cities that achieved a high status at the top of the urban hierarchy or were considered exemplars of urbanism were those that embraced new technology and became key sites for manufacturing and other industrial production. The successful city of the nineteenth and early-twentieth centuries was the industrial city, with cities like Manchester, Glasgow and Liverpool rising in importance. Symbols of success were manifest in the built environment and were actively used for city promotion purposes. Smokestacks,

22 Saltaire near Bradford. Though a 'philanthropic' community, this was one of the largest mills in nineteenth-century England. (Engraving.)

railways, bridges, factories and other industrial functions portrayed an area at the cutting edge of economic change (Figure 22). Entire regions, such as the Black Country in the British West Midlands, became characterized by a particular form of industrial activity. Cities began to specialize in various production activities with their very identities becoming intertwined with particular products; for example, Sheffield became synonymous with the steel industry and in later years, cities like Detroit and Turin became closely associated with the automobile industries. But while economic change had a tremendous and very obvious impact on the physical landscapes of cities through the construction of new places of work and production, so too did it have a dramatic impact on the social structure of the city. For the first time, a clear separation emerged between home and work; people began to commute or travel to their place of employment, albeit over relatively short distances; and direct personal connections between the employer and employee were lost as factory production gained in importance.

Throughout the twentieth century, this urban evolution continued, accelerated through the changing global context within which it was taking place.

Just as new technologies had facilitated the industrial development of European and North American cities in the nineteenth century, new technologies and in particular communication technologies have had dramatic impacts on cities throughout the world in the second half of the twentieth century. While the nineteenth century was undoubtedly one of internationalization and exploration, the late-twentieth has been the era of globalization. This phenomenon, driven by improvements in transport infrastructure, in particular the growth in the airline industry and in telecommunications, has radically altered the shape of and relationship between cities. While there have always been links between places located at significant distances from one another, the globalization processes we witness today are characterized by intensified links between people and places, and more rapid communication between institutions. Globalization has facilitated the greater mobility of individuals and organizations from place to place; events in one part of the world increasingly affect other areas at great distances; and, more often than not, decisions relating to investment, construction or other economic issues in specific places are made somewhere else. The general trend is a move away from transportation as the key determinant of success and growth towards an emphasis on the communication technologies that are driving the new service and knowledge economies. We can perhaps understand the change as a shift away from transporting goods to one of communicating ideas; the ship and truck has to a large extent given way to the laptop and internet.

Driving, but also taking advantage of, these changes are a range of economic activities that have had significant impacts on the shape of cities. Improved linkages at a global scale has meant that manufacturing, for example, can take place in one part of the world but be controlled from elsewhere. This so-called New International Division of Labour has had major impacts on Western cities from the 1960s onwards. The realization emerged that while cities in Europe and North America could remain as nodes for decision-making, this did not mean that all activities necessarily had to be carried out there. While headquarters functions of many companies have become concentrated in a small number of cities, particularly London, New York and Tokyo, identified as the key 'global cities' (Sassen, 1994), production or manufacturing has been dispersed, primarily to south-east Asia. The direct result, particularly in the mid-twentieth century, was that many cities in mainland Europe, northern Britain and the north-eastern United States were severely hit by production plants closing down, including car manufacturing in English cities such as Coventry (Jaguar) and Dagenham (Ford) and steel production in American

23 Abandoned warehouses in Liverpool, 1980s. (A. Parker.)

cities including Pittsburgh, giving rise to de-industrialization or the so-called rust-belt phenomenon. The facilities that had marked cities out in the nineteenth century as ascendant or cutting-edge cities, suddenly symbolized decline, economic stagnation and an inability to respond to change. Changing patterns of global economic development and trade resulted in major physical, social and environmental problems at the city-scale as inner cities became characterized by abandoned warehouses (Figure 23), high rates of unemployment and, in some cases, worrying levels of contamination and pollution, the legacy of previous industrial activity.

We can understand these changes as part of a broader set of transformations in society. As the overarching aim of most entrepreneurs is to make money, capitalism is continually evolving in search of a 'spatial fix', a place in which to invest and maximize profit (Harvey, 1978). If a particular location or activity ceases to generate sufficient profit, then capital may either switch sectors (e.g. to property speculation) or switch geographical location (e.g. from Western Europe to south-east Asia) in order to expand. This results in uneven development, particularly at the urban scale. The exploitative nature of this activity pits countries and cities against each other in the bid to attract investment, giving rise to 'place wars' (Haider, 1992) and sparking a whole new industry in urban marketing.

The impact on the urban waterfront

One of the reasons why the effects of globalization have been felt so severely in waterfront cities is that the movement from dependency on manufacturing and heavy industry to service industry resulted in the almost complete redundancy of traditional industrial areas, many of which were located in or close to port zones. This restructuring has had three key impacts, broadly defined as economic, social and spatial change.

The diversion of capital to emerging economies, such as Taiwan in the 1980s and 1990s and now increasingly to India and China, resulted in the degeneration of traditional industrial and manufacturing areas, including dockland areas, which struggled to adjust old infrastructure to new functions. Vacant piers and empty warehouses were the spatial manifestation of economic restructuring in traditional industrial areas of the city, creating serious problems at the time but also setting the scene for later capital re-investment and property speculation. The kinds of advancements underpinning the transformation of cities in the last fifty years have primarily been technological such as computer networking, internet services, and the development of rapid transport networks; traditional port activity was undermined even further by the simultaneous emergence of air transport, larger ships and containerization (Rybczynski, 2006).

Before the 1960s, large-scale international passenger movement was mostly facilitated by passenger ferry. In recent years, increased availability of air transport and a dramatic reduction in the cost of air travel, resulted in the loss of two of the port's primary functions – to act as a gateway through which people enter and leave a country and to operate as a node of importation. Yet even in ports like Dublin that successfully managed to grow port business, dockland areas became victims of the success of the port. Like the historic patterns that had forced ports to migrate seawards to deep water, the need for additional space for container storage and the new lifting cranes led to a move away from many dockland areas and the shift to more extensive sites; in many cases in areas that had been recently reclaimed from the sea.

In the late 1950s and early 1960s, it was considered normal for a ship to spend between 50 per cent and 80 per cent of time in port because of labour

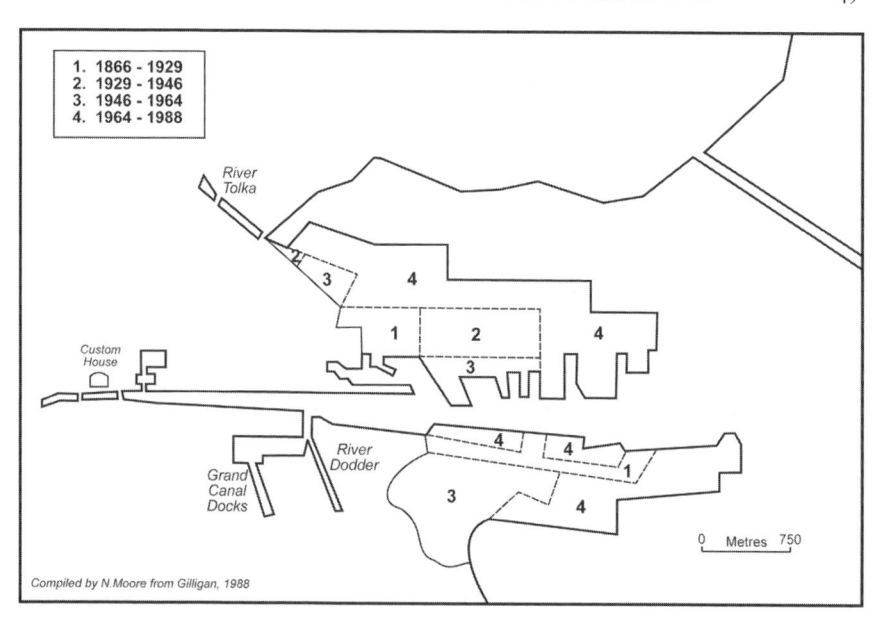

24 The extent of reclaimed land in Dublin port. (After Gilligan, 1988.)

25 Manual cargo handling at the B&I Terminal, North Wall Quay, Dublin. (DDDA.)

26 Container handling in Dublin port, 2003. (J. Brady.)

intensive cargo handling practice (Figure 25). In an increasingly competitive economic environment, this pattern became unsustainable and combined with changes in the type of goods transported and, in order to operate economies of scale, ship size grew rapidly. Immediate effects included the increased demand for even larger deep-water facilities as a serious mismatch between the provision of, and demand for, these facilities became evident. Few if any ports could support new technology without restructuring.

Combined with this shift, the widespread adoption of containerization in the 1960s has probably been the single greatest influence on the decline in dockland activity. Standardized containers are today a familiar feature in all port areas, but forty years ago they revolutionized the functions and character of ports (Figure 26). This new way of handling goods facilitated the easier and quicker movement of cargo impacting particularly on international trade. Combined with the introduction of other technologies, loading and offloading times of 4–10 days in Dublin in the 1960s have now been reduced to approximately 6–8 hours.

27 Docklands at Melbourne in 1997 as renovation begins.

This revolution in the relationship between ports and their cities has had a global impact. In Melbourne, for example, nineteenth-century classifications of the docklands as a wasteland attracted industries such as abattoirs, candle makers and wool washers and the expansion of these, Melbourne's first industries, resulted in the downstream expansion of wharves and piers. Like most waterfront areas, subsequent developments resulted in transport and manufacturing becoming the primary land uses, with old gasworks piercing the skyline in the Victoria Harbour precinct. Changing technology and transport requirements resulted in the construction of berths downstream at Victoria Dock (1890s) and at Appleton and Swanson Docks (mid-twentieth century) to facilitate large ocean-going vessels, in much the same way as Dublin port, at the same time, advanced seawards and activity moved to the new deep-water Alexandra Basin, discussed in the previous chapter.

Technological change and the growing importance of air and road transport in recent decades resulted in the closure of some docks in Melbourne particularly the easterly wharves, which were decommissioned in 1975. In Dublin docklands of only eighty years ago, abattoirs and chemical factories were located cheek by jowl with the residences of those dependent on port

28 A view down the Liffey in the early-twentieth century. (Postcard.)

29 A view down the Liffey in 1976. (J. Brady.)

activity for survival. When these facilities closed, their legacy contributed to the widespread negative perception and marginalization of the docklands area that had become widely engrained in the psyche of Dubliners and others by the early 1980s.

An international stereotype? Dublin in the early 1980s

Less than thirty years after the advent of containerization in the port, Dublin docklands was a very different place to the bustling trading port of the early-twentieth century (Figures 28 and 29).

Traditionally a heavy industrial area of the city, the economy and employment base of docklands declined sharply from the late 1950s. A combination of factors including those mentioned above resulted in economic, physical and social decay. Government policy, particularly the Wright Plan commissioned by the Minister for Local Government in 1964 though never formally adopted, advocated the development of new towns in the outer suburbs. This did little to alleviate the outward migration of activity and resources and was partially responsible for the abandonment of the inner city. As suburbs to the north, west and south of the city expanded rapidly, the inner city hollowed out. Between 1971 and 1991, the population of inner city Dublin fell from 567,866 to 478,389 (Brady, 1988; Census of Population of Ireland, 1991). As a major part of the inner city, the waterfront areas experienced all the symptoms traditionally associated with inner urban decline, including falling population numbers, the loss of industrial employment, a poorly skilled workforce, high rates of unemployment and vacant buildings (MacLaran, 1993). In both the north-and south-eastern sectors of the city, the traditional emphasis on marine-related employment and a history of low educational attainment resulted in chronic skill shortages for non-manual employment. Combined with the lack of investment in private housing, an increase in the number of vacant sites and local authority housing policy, physical and social dereliction accelerated rapidly (Figures 30 and 31). The contrast between the ways of life in the heyday of the docks with that once decline set in is remarkable:

> I'd say there were twenty shops on Sheriff Street. You'd get it all. You'd see them coming down on Saturday evening with their vegetables and all for Sunday's dinner. There's nothing in it now. A pork butcher's and a pub.
>
> (North Inner City Folklore Project, 1995, p. 12.)

30 Dereliction, Summerhill, 1983. (J. Brady.)

31 Dereliction, Mountjoy Square, 1983. (J. Brady.)

32 Aerial photograph of Sheriff Street flats, 1994. (DDDA.)

Local history projects undertaken within the docklands have recorded the dramatic changes brought on by changes in port activity experienced by the communities north and south of the river:

> It was a fantastic sight to see so many men in the darkness of the morning going down to a little wooden boat to go across the river to get work on the Dublin Docks. As it declined at the latter end of the war ... I was very disturbed to see a large number of these dockers getting the same little wooden boat back again due to the fact that there was no work for them. They started off in small numbers of twenty to one hundred. Then it went into hundreds, getting turned away for work, very depressing it was.
> (St Andrew's Heritage Group, 1992, p. 56.)

The docklands in Dublin, which had been among the worst nineteenth-century slum areas, had come full circle to become some of the most destitute districts in the twentieth-century State (Prunty, 1988). The most striking feature of the area in the early 1980s was the lack of open space and recreational amenities in the area. Major land uses included transport, energy/

33 Local authority housing at Thorncastle Street, Ringsend. (N. Moore.)

utilities, industrial and wholesale functions; one might expect to see these activities in districts on the periphery of cities away from residential areas. On the north side of the river Liffey, the main concentration of residential use was the Sheriff Street/East Wall/North Strand area, while on the southern side Pearse Street and Ringsend were the focal points. The largest residential complex in the vicinity was the Sheriff Street flat scheme, constructed between 1930 and 1952. The 18 four-storey blocks constructed to house the large numbers of dockers resident in the area consisted of 445 individual units comprising two- or three-bedroomed flats. But while the flats were initially constructed to house dockers, the decline in that kind of employment resulted in a large number of families moving out. The local authority later allocated vacant flats to families at risk from other parts of the city, but this policy combined with a range of other factors, including lack of employment opportunities, led to the rapid physical and social decline of the area (Figure 32). The unbalanced social profile of the area was compounded by the lack of private sector investment in the inner city and the concentration on housing development in the suburbs. Even so, very successful and indeed popular housing complexes were also constructed in the 1950s, such as that at

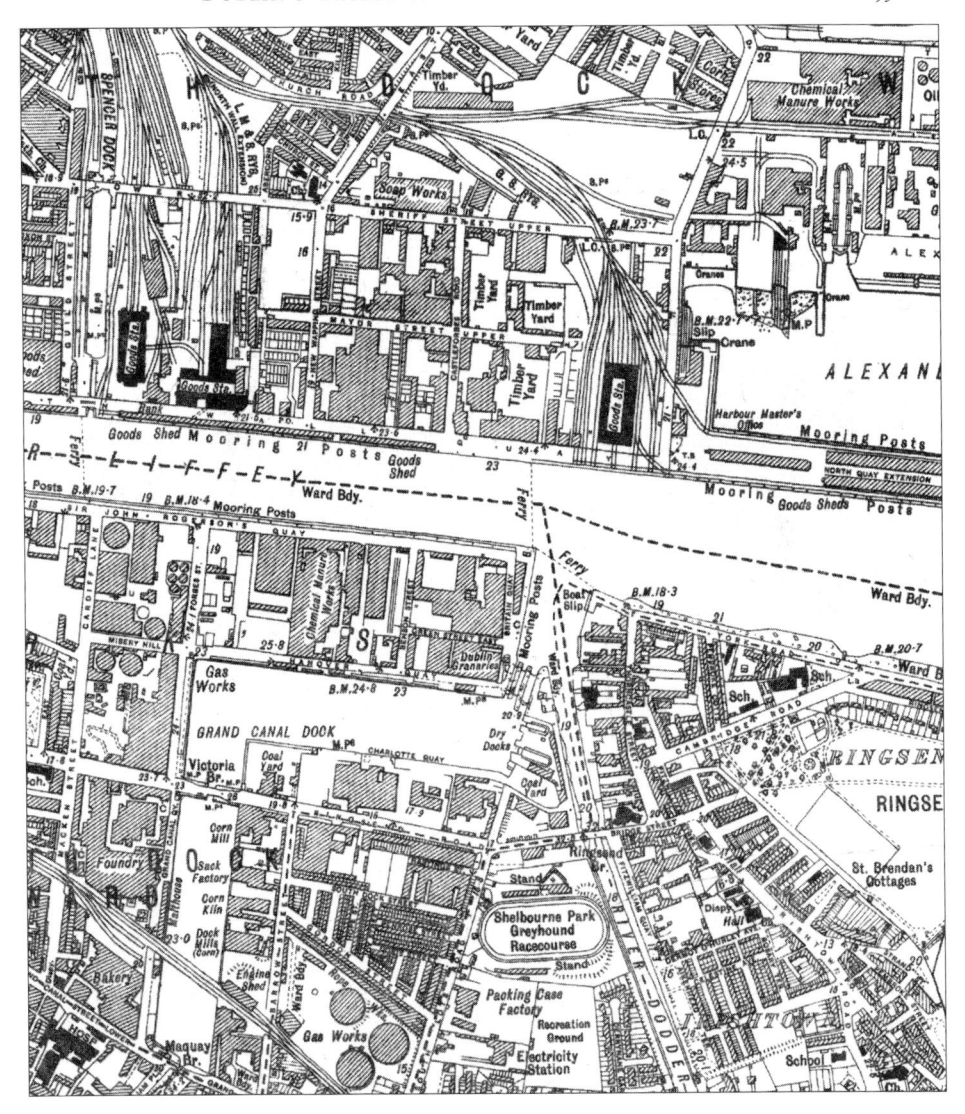

34 Industrial and residential landuses in close proximity.
(Ordnance Survey 1:10,560 plan, Sheet 18, Dublin, 1938 revision.)
Note the Barrow Street Gasometers, Spencer Dock railyards and the chemical works.

Thorncastle Street in the southern docks, still an important component of the city housing stock today (Figure 33).

In contrast to the city suburbs that had an emphasis on individual dwellings in a leafy green setting, in both Sheriff Street in the north and Thorncastle Street in the south docklands, heavy industrial zonings surrounded the residential

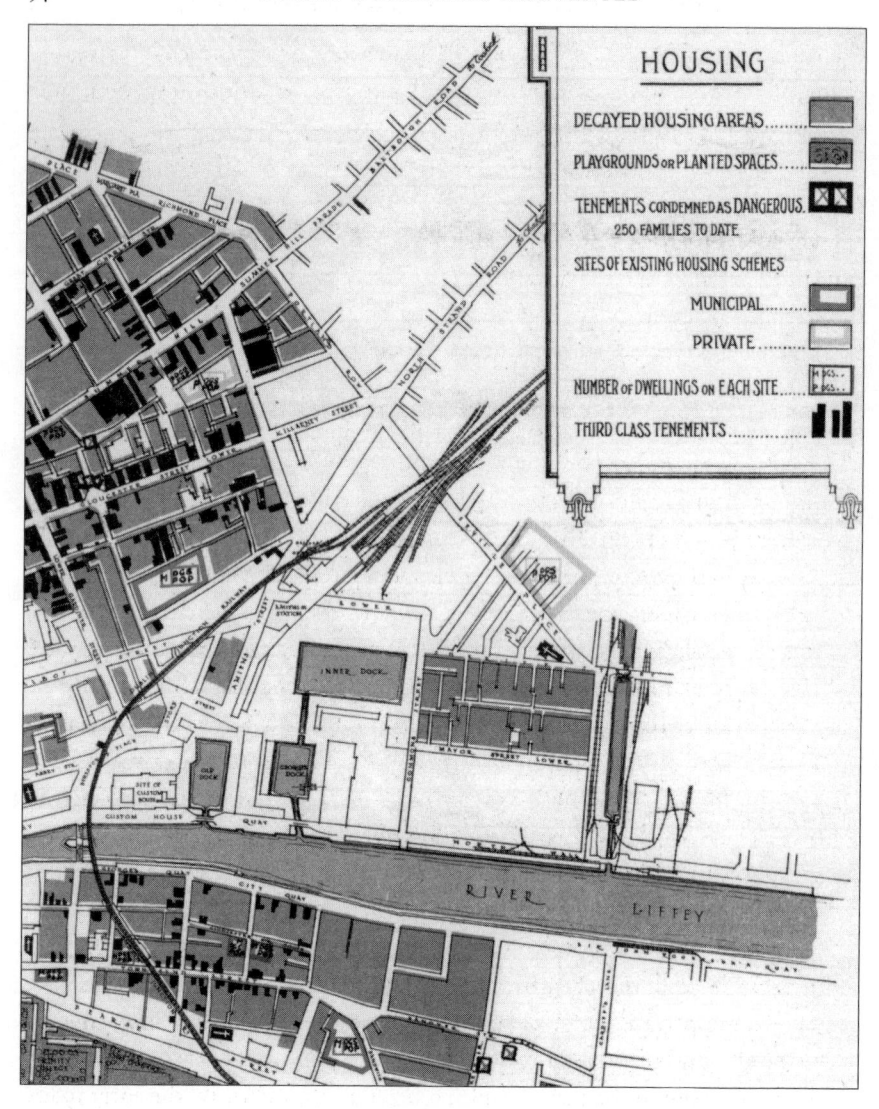

35 Housing quality in Dublin Docklands. (Civic Survey, 1925.)

areas. The gasometers were the most significant feature in the landscape south of the river and the railway sidings and container yards encircled homes on the north (Figure 34). In the Grand Canal Dock and Ringsend areas, warehousing, transport and industrial activities were particularly dominant. It is no surprise, then, that with a global move away from manufacturing industry towards the service economy these areas found it particularly difficult to make the transition. What had once defined these areas as economically successful eventually proved to be an albatross that prevented, or at least impeded, their modernization and transformation.

Coupled with physical challenges, the selective nature of out-migration to the suburbs exacerbated existing social problems within the urban core. The low skill levels of 'residual' populations have been recognized as both a symptom and an effect of general inner-city decline. Dublin proved no different in the early 1980s. Labour de-casualization and reductions, resulting from increased mechanization and the introduction of new work practices at Dublin port, meant that for many, traditionally dependent on port, manufacturing and industrial activities in the north-east inner city, the only avenue to employment had been removed. In the nine-year period 1975–84, employment at Dublin port was reduced from 7,403 to 5,200. From 1981 to 1986 unemployment in Dublin County Borough (now known as 'Dublin City') increased by 91.6 per cent (Dublin Corporation, 1986). An unemployment rate of 24 per cent in the north inner city was recorded in the 1981 Census of Population; 30 per cent of the unemployed had been without work for five years or more, significantly reducing the chances of becoming re-employed; 52 per cent of heads of household in Sheriff Street, in the heart of docklands, were on the unemployment register, a figure that had increased to 70 per cent by 1986 (Dublin Corporation, 1986). A survey commissioned by Sinn Féin/The Workers' Party in 1982 estimated that in the City Quay/Pearse Street area in the south docklands, over 2,700 jobs were lost between 1971 and 1981 (*Sunday Press*, 25 April 1982).

The overall result was poverty and dereliction in parts of the docklands on a much larger scale than in other parts of Dublin. Housing conditions were very poor, the private sector having failed to provide any form of residential accommodation in the area for many years and local authority budgets being totally inadequate to the task (Dublin Corporation, 1986). By the early 1980s, over 50 per cent of housing in the docklands was over seventy years old and of very poor quality. The same trend of inadequate housing that had been identified in the North City Survey of 1919 and the Civic Survey of 1925, both of which had recommended wide scale housing demolition in the area, remained evident in the 1980s (Figure 35).

Conditions in one part of the south docks only twenty five years ago were more reminiscent of a third world city than the capital city of a European State:

> A family of seven tries to sleep cramped into one bedroom of their flat ... Elsewhere six children in a two-bedroomed flat queue to wash in a small blue plastic bathtub in the tiny kitchen of their home. There is no bathroom. And in another terrace of ageing flats a mother provides meals for her family on a cooker that is wedged into the toilet.
>
> (*Sunday Press*, 25 April 1982.)

Yet today many of these same areas have become highly desirable and sought-after locations in which to live with property prices exceeding many other parts of the city. While some of the same residents remain, a whole generation of new, upwardly mobile residents has been attracted to the area. Why this turnaround has been so dramatic and rapid is a complex issue, but part of the answer lies in changing global urban development and management trends. While the capital city in the 1970s mirrored to a significant extent the kind of decline that was being experienced in many other cities, by the end of the century it was beginning to emulate the strategies and ideas that were reaping dividends, at least for some groups, in places like London, Baltimore and Rotterdam. While Dublin can be regarded as stereotypical of general global urban decline in the late-1960s through to the early-1980s, it is no less true to consider transformations in the city in the last twenty years as emblematic of wider global processes of urban revitalization. This has been driven by the move away from the conception and operation of cities primarily as sites of production towards an emphasis on the development of the knowledge economy and the creation of post-industrial spaces, emphasizing consumption, spectacle and a particular type of city image.

Responding to change: the waterfront as urban playground?

One of the greatest effects of restructuring and indeed of the decline in cities up to the early 1980s was the dawning realization that cities, now operating in a more open global environment, needed to become more competitive and proactive in attracting new functions and investment. The dramatic increase in the mobility and type of capital circulating around the globe has provided new challenges and opportunities to cities. The target of their attention has now moved beyond simply attracting footloose hi-tech industries to embrace the economic potential of housing global headquarters functions for agencies including the UN, World Bank, World Health Organization, and hosting significant global events including the Olympics, World Cup, and major business conventions. But, because capital is more mobile, cities that were once the focus of industrial activity now face stiff competition from emerging cities, such as Atlanta, that are unconstrained by the legacies of previous activities and can embrace the new economy without any major structural adjustments. Intensified competition has spawned not only new global players in terms of urban activities but also new ways of doing business.

Local authorities have had to move away from the 'managerial' or development control approach that characterized their operation in the 1960s and 1970s towards a more 'entrepreneurial' or innovative approach to city management and marketing. Three strands have been identified in this approach: the first is the reliance on the concept of public-private partnerships (PPP). This is particularly evident in dockland areas where the majority of redevelopment projects have been undertaken by specially established development agencies such as the London Docklands Development Corporation or the Melbourne Docklands Authority working closely with private sector investors. The second characteristic of this approach is its speculative nature as the role of the public sector in these projects has in most cases absorbed all the risk associated with the projects to satisfy the private investors. Thirdly, these new developments tend to be place-specific rather than jurisdictional, focusing on particular projects rather than on redeveloping an entire local authority area (Harvey, 1989).

However, the only way this 'entrepreneurial' approach can succeed in attracting appropriate levels of investment is through the creation of a 'good business environment'. At a time of increasing inter-urban competition this generally means the promotion of a 'quality living' as well as working environment. To maintain their position within the urban hierarchy, cities must promote themselves as attractive places in which to work, rest and play, resulting in the inclusion of professional marketers and consultants in the revitalization process. Understanding city-image building or 're-imaging' is the key to comprehending the dynamics of city development and management in the twenty-first century, as the city is now, like any other commodity, being marketed and sold. As well as generating actual change, the role of many of these new agents of urban change, is to alter the outside perception of particular places. One of the difficulties indeed in reversing earlier dereliction in former industrial cities has been to persuade potential investors that the negative images they may hold of particular places no longer apply. There are a variety of ways in which this is undertaken, but many very simple techniques are used including the adoption of city logos and slogans. One of the most evocative slogans used in an urban marketing exercize has been the 'Pittsburgh: Stronger than Steel' logo adopted in Pennsylvania. The aim of all of this activity is to neutralize the negative and elevate the positive images of a particular place. The perception of image can be manipulated in an aggressive campaign without substantial change actually being made and local authorities have adopted a hitherto ignored role in city marketing and promotion. These tend to focus on the creation of images of service industry, business, lifestyle and environment. Successful marketing might include reference to the highly designed, green landscapes of the science park; the promotion of 'flagship' office buildings or activities that emphasize links between business and education; the idea of the city as a location not only for business but also for living centred around cultural facilities such as museums, exhibition centres and opera houses; and the reinvention of the urban environment as a collage of positive images including spectacular and historic architecture. Nonetheless, the marketing rings hollow if these facilities fail to be delivered, so it is no surprise that all of these elements feature very strongly in the kinds of regeneration and spatial restructuring promoted in many dockland projects.

In waterfront areas, land use change is the most visible impact of redevelopment. The majority of former docklands have been, or are in the process of being, transformed from industrial transport-oriented sectors to mixed-use developments. The emergence of a 'standard' waterfront package or formula for successful redevelopment, drawing on the images of business, lifestyle and environment, appears to have transcended all political and cultural boundaries, with Baltimore and Sydney being prime examples. Sydney's Darling Harbour revitalization scheme came on stream just as Baltimore's first renaissance was nearing completion in the mid-1980s, yet the two projects are uncannily similar, exhibiting the same range of elements including festival marketplaces, museums, restaurants, large office developments and high-class residential property. In addition to the construction of new buildings, warehouse conversions have been a characteristic of waterfront residential renewal and are among the most prestigious properties in these areas today aimed at young professionals. The resultant gentrification has been the subject of controversy in many places, most notably at St Katharine's Dock in London. The result has been the marginalization of traditional communities within the renewed areas, and the displacement of dockland residents, a general criticism of urban regeneration projects in many cities. This kind of development seems to be part of a new social organization within cities trying to adjust to the new global economy, the new patterns being described as:

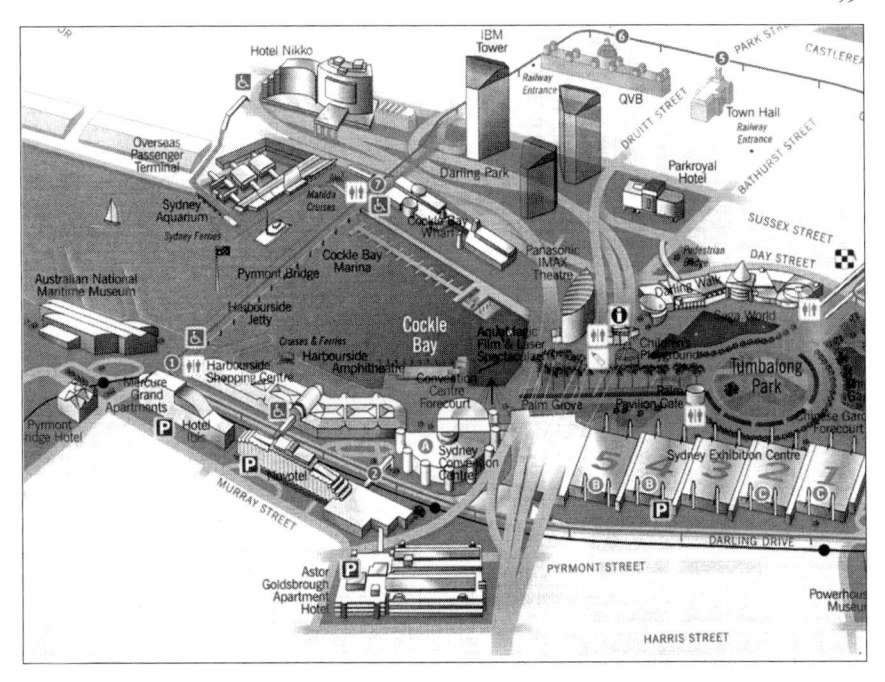

36 Outline diagram of redevelopment, Sydney. (Sydney Harbour Authority.)

37 Darling Harbour, Sydney. (Sydney Harbour Authority.)

38 Darling Harbour after redevelopment in 1998. (Ian Elliott.)

a spatial concentration within cities of a new urban poverty on the one hand, and of specialized 'high-level' internationally connected business activities on the other, with increasing spatial divisions not only between each of them but also among segments of the 'middle class' in between ... Boundaries between divisions, reflected in social or physical walls among them, are increasing. The result is a pattern of separate clusters of residential space, creating protective citadels and enclaves on the one side and containing ghettos on the other.

(Markuse and vanKempen, 2000, p. 3.)

It would not be fair to suggest, though, that this is an inevitable trade-off for development as this kind of outcome has been deliberately avoided in some places including the Inner Harbour redevelopment in Baltimore, Maryland. As we will see later, while the situation in the early years of redevelopment in Dublin could be characterized as conforming generally to the above pattern, the relevant agencies in recent years have attempted to avoid this kind of outcome.

Perhaps a more striking and defining characteristics of regeneration has been the attempt to draw heavily on the lifestyle dimension of redevelopment. Rather than simply reinvigorating city centres through the attraction of residents back from the suburbs into new buildings, a much wider range of people including day-trippers, tourists and business travellers are targeted through an emphasis on creating recreational 'destinations', promoting the idea of old dockland areas as new urban playgrounds focused on the consumption of leisure time activities. In Sydney, for example, the city authorities adopted a 'theme park' approach to regeneration introducing a new convention centre, marina, festival marketplace and hotels costing over AUS$900 million (Figure 36). The harbour is now not just renowned for the Opera House but is ringed by an aquarium, waterfront promenades – Darling Walk and Cockle Bay Promenade – a maritime museum and small parks (Figure 38). The Chinese Garden of Friendship and Tumbalong Park are the major public space developments, the former having evolved in recognition of, and one could perhaps argue to encourage further, close commercial and community ties between China and Australia. The construction of a ceramic Dragon Wall within the garden is symbolic of the ties between the two countries. Sega World's $70 million indoor game park and an IMAX theatre featuring some of the largest screens in the world have been recent additions to the recreational landscape. One of the most significant individual projects, Cockle Bay Wharf, was billed by developers during construction as an exciting, flagship development providing a whole entertainment package: three levels of dining for visitors, a nightclub and upmarket restaurant.

While Cockle Bay Wharf was one of the first such purpose-built precincts developed in this kind of location, the same strategy is now evident in other places such as the Port Olimpico in Barcelona and the Odyssey Centre in Belfast. Further supporting the idea that global economic shifts have resulted in the transformation of these districts from sites of production to those of consumption, is the importance afforded to retailing as a lifestyle activity. In an age where consumption, and indeed conspicuous consumption, has become a key leisure-time activity, retailing has not only become a significant economic function but also plays a key role in ensuring the vibrancy of new areas by acting as a magnet or attraction to different groups, particularly to 'festival marketplaces'. First developed by the Rouse Company in the USA, a typical festival marketplace generally offers major restaurants, specialty and trendy retail shops, and an international food court. The most successful of these marketplaces have resulted in transforming areas that were once plagued

39 Bristol Harbour, the Arnolfini centre for contemporary arts. (N. Moore.)

40 South Street Seaport, New York. (D. Hockman.)

by de-industrialization to post-industrial spaces of consumption, such as Station Square in Pittsburgh and Waterside in Norfolk, Virginia. The very first festival marketplace, Harbourside Shopping Centre, was developed at the Inner Harbour in Baltimore as a series of two storey pavilions linked by a central plaza, providing a mix of retail outlets and eating-houses. The overall aim of creating these kinds of facilities is to attract global investment capital by offering a unique blend of work, leisure and residential services; but the strategy is itself being replicated globally in places as far removed as New York's South Street Seaport and the V&A Waterfront in Cape Town, thus negating its purpose and intensifying competition further.

Clearly, globalization has put pressure on cities to develop their specific cultures and environments in ways that attract business, investment and high-tech professionals and that convince their own residents and entrepreneurs to remain. Yet in many cases what has occurred is that diverse cities are becoming more similar as they all try to emulate similar ideas and promote a similar conception of the kind of product that might be successful on a global stage, ultimately negating marketing strategies and intensifying competition even further. Undoubtedly, re-imaging and actual physical change have been successful in reversing the fortunes of many waterfront cities and led to the rediscovery of waterfronts as urban playgrounds (Figures 39 and 40), but a variety of important issues remains to be addressed.

In cities such as Toronto, access to the waterfront has been constrained as redevelopment has *de facto* privatized whole stretches of the lakeshore through the construction of private residences. This is an important issue in an Irish context, as recent proposals for residential development in Dun Laoghaire, Dublin, would have a similar impact. Other concerns generated by the kinds of waterfront regeneration that have been promoted globally centres on their role in re-integrating the port with the city. In many cases, the new projects have failed to generate any spin-off benefits to other areas within the urban core. It has been argued that the ethos behind redevelopment is inherently inward looking, even sub-urban or anti-urban, because these areas have been developed in a vacuum without examining their impact or relationship to other areas (Malone, 1996).

Without doubt, globalization has resulted in greater social polarization within cities. Rather than narrowing the gap between the areas that were derelict and the rest of the city, in many ways the untold economic success of these areas has made life more difficult for many. While new global enterprizes have generated some high-powered and elite employment, there has been a significant expansion in the number of very low paid jobs servicing these industries and a significant growth in the informal economy. In fact, the impact of large-scale redevelopment projects has been to foster 'dual cities', creating some winners but many more losers (Logan and Molotch, 1987). While local agencies have responded to the urban challenges of globalization by focusing on physical and economic renewal, many social problems generated by deindustrialization and economic restructuring have been ignored or in some cases intensified. This was a criticism levelled at early redevelopment in London Docklands, the project becoming a symbol of all that was wrong with the neo-liberal economic policies of Margaret Thatcher and epitomizing the divisions that were splitting society down the middle. Despite the economic boom in the London Docklands in the early 1980s, unemployment persisted at unusually high levels among younger residents. When average income in the area was £8,500, new two-bedroom apartments in the area were priced at £185,000 (Short, 1996). The continuous pricing of housing at these levels occurred concurrently with an 81 per cent rise in homelessness, from 1979 to 1986, within the Docklands boroughs as a whole, negating the idea that spin-off benefits from new developments would help older areas. The attraction of skilled service and financial employment to the area did little to benefit a workforce traditionally dependent on manual activities. In the period 1982–87, unemployment within the London Docklands Development Corporation area increased by over twice the average Greater London rate (Kyne, 1989). While a large number of jobs were created, they were aimed at those who commuted to work each day from locations outside Docklands rather than the residents of the area that had been most adversely affected by industrial restructuring and port decline.

This is typical of the way in which modern urban life is changing and indeed becoming increasingly diverse and difficult to manage. In earlier decades, urban space was organized so that different social groups occupied distinct spaces within the city. It would have been rare to see the poor and the very wealthy living cheek by jowl in most cities in the late-nineteenth and early-twentieth centuries, and even the development of suburban housing areas traditionally had a very specific housing profile. Because of the very narrow social structure of residents and visitors that redeveloped waterfront areas have attracted, primarily young, unmarried professionals with significant levels of disposable income, the profile of these areas that were traditionally working-class and dependent on manual labour, has become much more complex. While the cultural or social distance between social groups is continuing to increase in cities throughout the Western world and levels of polarization grow, the physical distance between diverse groups has decreased rapidly. This has caused significant tension and resentment between different groups within the city and raises issues around whether a coherent community will, or can, ever emerge in waterfront locations. Dramatic and unprecedented in terms of its scale and intensity, the transformation in cities over the last twenty-five years is nonetheless just part of much longer term processes in urban evolution. The landscapes we see today represent the outcome of the most recent political and economic decisions taken by a range of 'actors' or agents who have wielded power in our cities. While particular regeneration projects might be characterized as 'the product of a crude and opportunistic capitalism' (Malone, 1996, p. 77) operating on a global scale, they are still undertaken within specific national and local contexts. Local agents and decision-makers still play a key role in determining how individual projects are conceived and delivered, facilitating or constraining the ability of global capital to achieve that ever elusive 'spatial fix'. This was particularly evident in Dublin in the 1980s where national political needs provided the first impetus to waterfront regeneration in one of the most challenged districts within the urban core.

The politics of planning docklands, 1980–1987

> For those of us who were born here, the sad reality is that we can no longer take pride in calling ourselves Dubliners. The state of the city is a cause for shame, a cause for outrage. Even our City Fathers are prepared to concede that Dublin is rotting at the core.
>
> (McDonald, 1985, p. 2.)

Given the difficult economic climate in Ireland, and indeed globally, in the late 1970s, it is no surprise that by the early 1980s so many towns had become characterized by dereliction and decay. How governments would, or could, respond became an increasingly difficult question as the needs of local urban communities almost seemed to run at odds to what was emerging as best international practice for promoting future economic growth. In Ireland, the traditional role of the local authority had always focused on development control. In a context where very little development was taking place this role became increasingly irrelevant and the role of the local authority began to be questioned. At the present time in Ireland there is some debate about the most appropriate relationship that should exist between the planning and political systems in the light of evidence that has been unearthed by Tribunals of Inquiry into planning corruption. In public debates, the Fine Gael TD, John Deasy, has called for a complete separation between politics and the planning process, blaming the close relationship between the two systems for the lack of comprehensive, rational planning that has characterized Ireland in recent years: 'The reason there is inconsistency in the entire planning process and houses are being built on skylines throughout the county is that councillors for the last 20 or 25 years throughout this country ... have pressurized planners and county managers' (*Irish Times*, 17 March 2003). In the depressed economic climate of Ireland in the 1980s, it was not just local councillors who were closely involved in local planning and development issues but also the key players in central government. While there is reason to question how these relationships operated at that time and the current planning tribunals clearly demonstrate this fact, it is not possible to achieve a total separation of the planning and political systems. Because spatial planning has the redistribution of people, wealth and resources at its heart it is an inherently political activity.

The broad framework within which planning operates is established by political powers and any decision-making will more often than not generate winners and losers. The planning system shapes development and this is itself shaped by political ideology and the economic environment. Thus to view planning in a positive light, as opposed to the more cynical view that many Irish taxpayers perhaps rightly hold having heard repeated reports of corrupt planning decision-making for the last decade, would be to argue that it mediates the relationship between politics, the economic system and the landscape. As was argued in earlier sections dealing with the context for redevelopment, since the 1980s the social, economic and political environments in Ireland and elsewhere have been subject to wide-ranging restructuring. As a direct result, planning policy has also changed quite substantially. The transformation might, at least in theory, be characterized as a move away from a system of prescriptive development control to one of development promotion with some collaborative ventures drawing on the social partnership model dominant in Ireland. But, as in all things, the reality is much more complex and has been so, particularly in the old port or docklands part of Dublin, for at least the last twenty-five years.

The Irish planning framework

As in the United Kingdom prior to the passage of the 1947 Town and Country Planning Act, early Irish planning legislation lacked any kind of strategic element, and planning for the most part was reactive in nature. Combined with a changing socio-economic climate, generated in part by a re-orientation of economic policy from protectionism to free trade, this approach produced difficulties in terms of managing growth, particularly in the capital where the population of Dublin had expanded by almost 16 per cent, from 718,332 in 1961 to 852,219 in 1971 (Census of Population). This level of growth in one decade was equivalent to the total population increase that had occurred over the previous forty years, and resulted in a significant increase in the spatial extent of the city.

Re-assessment of planning policy was an imperative, and the Government responded by inviting Charles Abrams to Ireland to help devise a rational and efficient planning system. Abrams, who spent his life working on housing reform, urban planning and the links between the public and private sector, was both an academic and a policy-shaper. He worked on United Nations missions in the developing world, playing a critical role in reconstruction work in Ghana (1954), Nigeria (1962), and Kenya (1964), but was also Chair of the Planning Department at Columbia University, New York. Bringing his international experience to Ireland, Abrams' final report recommended that a new planning act should be introduced to function in a similar manner to the British Town and Country Planning Act, 1947. In 1963, the landmark Local Government (Planning and Development) Act passed all stages of the Oireachtas, creating a completely new planning environment. Section 3.1 of the Act defined development as encompassing both new construction and the alteration of land use, which would be managed through strategic land-use zoning for large areas, combined with a requirement for individual planning permissions for specific proposals. In contrast to previous legislation, the 1963 Act was a statutory instrument and therefore had a strong legal basis with powers of enforcement. But the major difference between this and the type of legislation that had come before, and been totally ineffective, was that the production of a development plan, 'a statement of planned policies for the local community and the basis on which all development proposals are considered' (ENFO, September 1996, p.1), became mandatory.

While being a dramatic improvement on previous legislation, the Act was still fundamentally flawed, the implications of which are being felt today. The lack of any references to regional, or the co-ordination of physical and other, planning resulted in large urban areas like Dublin being managed by a plethora of different agencies each with their own priorities. The result has been inefficiency, planning chaos and until recently, a sometimes unfavourable institutional environment for the private investor. The Myles Wright plan of the 1960s proposing the development of four new towns on the western edge of Dublin separated by a greenbelt was the first serious attempt to think in a strategic manner, and has only recently been updated by the introduction of Strategic Planning Guidelines for the Greater Dublin Area. Introduced in 1999, these aspired to giving some coherence to the planning of Dublin and the mid-East region. The guidelines provide a framework for planning coordination between Dublin City, Dun Laoghaire–Rathdown, Fingal, Kildare, Meath, South Dublin and Wicklow County Councils. The publication of the National Spatial Strategy in December 2002 has also tried to outline an overall strategic national framework within which development will occur. This twenty-year planning framework for the entire country aims to achieve a more even regional balance across Ireland in terms of social, physical and economic

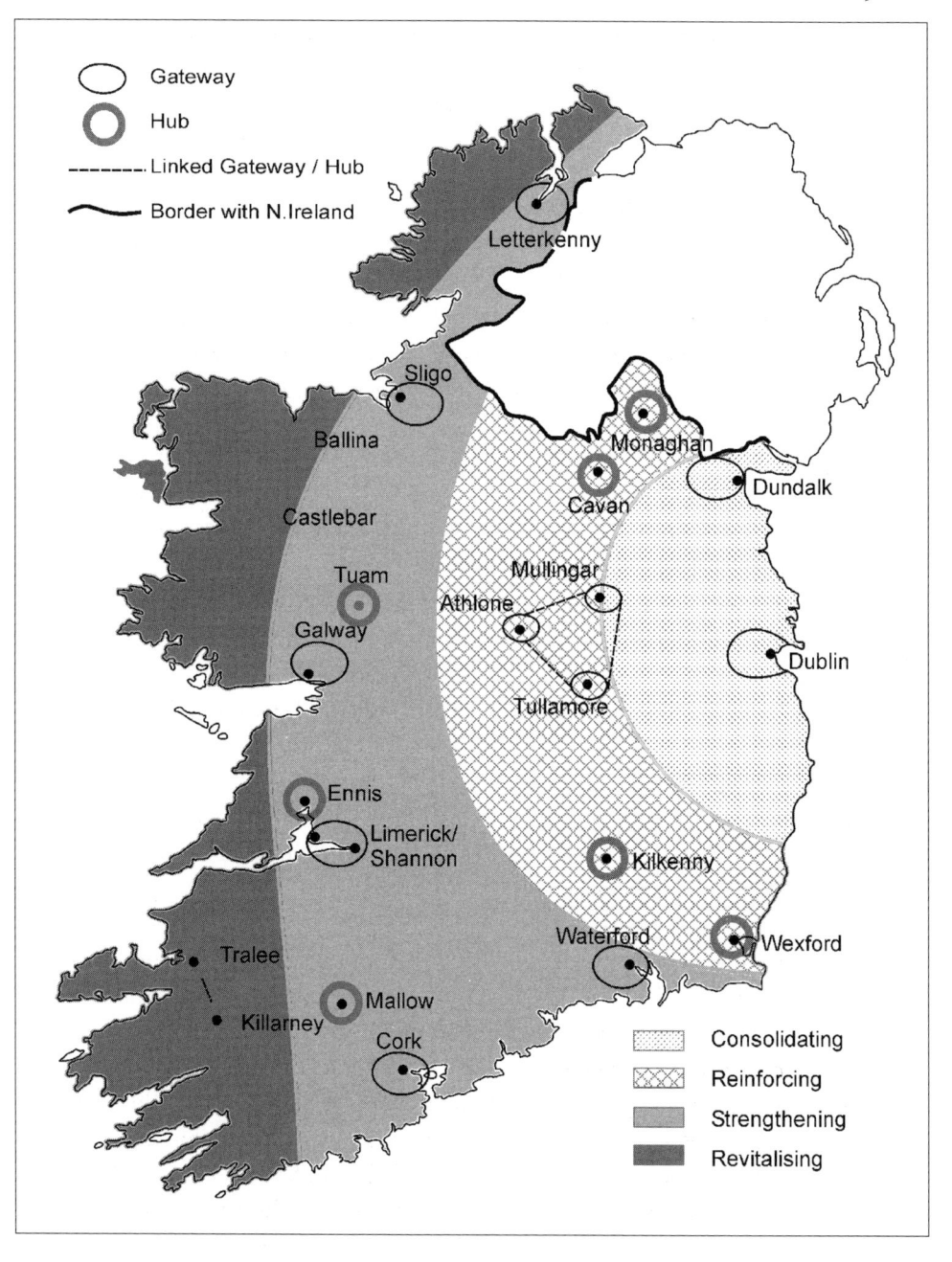

41 Gateways and hubs in the National Spatial Strategy. (Redrawn from NSS, 2002.)

development. Existing major cities – Dublin, Cork, Limerick, Galway and Waterford – will continue to function as 'gateways' or engines of regional and national growth, while other key urban areas where new growth will be concentrated have been defined as 'hubs'. Understandably, the publication of the document became politicized with some local organizations, such as Carlow Chamber of Commerce, highly disgruntled when their district was omitted from the list of favoured areas in favour of neighbouring towns such as Kilkenny (Figure 41). The programme of government decentralization introduced in the Budget in December 2003 has again highlighted the problems that political decision-making creates for effective planning as many of the locations identified as new government departmental headquarters are not those targeted as growth centres by the National Spatial Strategy. Many newspaper reports and commentators have argued that decisions regarding the future location of government offices and thus the prospect for development in particular areas has been driven more by an attempt to gain votes rather than any over-riding strategic spatial objectives.

Returning to the discussion of the general planning environment in Ireland, another major difficulty with the 1963 legislation – that remained the key planning document until 2001 – was the heavy emphasis placed on regulation rather than development. Local authorities were granted restrictive rather than permissive powers in contrast to North American planning legislation, which tended to favour growth. Although regarded as a planning and development Act, the 1963 legislation placed undue emphasis on control to the neglect of development aspects. This predominantly negative approach in Irish planning practice was very similar to that embraced by the British planning system (Guy, 1994). In recent years this criticism has become less valid as local authorities evolve into more entrepreneurial and pro-active agencies, taking on the key functions of attracting investment and promoting economic development as will become clear later in this, and in subsequent, chapters.

In terms of the political, rather than just the pragmatic change that the 1963 Act brought about, a public role and mechanism for participation was enshrined for the first time in legislation. Every non-exempted development would require an application for planning permission to be lodged with the relevant authority and public consultation to be undertaken. While the public can influence the process, recent criticism has centred on the apparent inability of the system to meet the demands of a well-educated public for a more constructive participatory role (Bannon, 1989). Many writers including McGuirk (1995) suggest that the Irish system has promoted a negative role for public participation arguing that, only after planners have decided on their preferred options is the public provided an opportunity to comment. She ascribes the absence of public participation to the attitude of planners: 'one planner conceded that "(planners) are expert enough to make (planning) complex enough for the average person not to understand" ' (McGuirk, 1995, p. 72). This has been undermined even further by the power vested in city managers to override decisions of the democratically elected representatives.

While not political in the electoral or voting sense, this kind of attitude is inherently political in a broader sense as it generates inequities in terms of the ability to influence decision-making, rendering some groups more powerful than others. But what is so interesting about the story of docklands, is that 'politics' in all senses of the word has become critical to the way in which the redevelopment project has evolved from as early as 1980.

Redevelopment plans of the Dublin Port and Docks Board

Given the changing nature of port activities both globally and locally and the shifting location of the maritime function within Dublin city following mechanization and the move towards containerization, the role of former port lands was becoming a matter of increasing concern for the Dublin Port and Docks Board by the early 1970s. In late 1972, a comprehensive document outlining both the physical and economic development desired by the port authority over the following ten years was published, including proposals to reclaim 2,000 acres (809 ha) of Dublin Bay, provide a welcome 8,000 new jobs and increase car-ferry traffic by 150,000 by 1980. While undoubtedly ambitious, the proposals resulted in much disquiet given that they ran completely contrary to the Government's previously stated objectives of not actively stimulating growth in Dublin (Press release, 19 May 1969) and distributing industrial development throughout the country. The plans envisaged developing the Custom House Docks as a limited mixed-use site with some industry, including metal processing and chemical works, along with residential development of over 20,000 new homes. This kind of juxtaposition would not have been unusual within the context of existing land use mixes along the waterfront, but the active planning of incompatible land uses adjacent to one another came under fire by many organizations, including An Taisce (Figure 42). They argued that acceptance and support of these plans, outside the regular planning and development channels, would imply that both the port and local authorities were planning agencies, albeit operating

42 An Taisce publication opposing the port proposals, 1973.

polar urban policies. The Sandymount and Merrion Residents Association who were vehemently opposed to the port plans argued that the Port and Docks Board:

> has no right to spend public money on promoting grandiose schemes which have nothing to do with port facilities and which are completely incompatible with the recommendations of the responsible planning authorities and the Government ... All this, at the public expense, to persuade the public that a 300 yard wide strip of land separating the sea from the petrochemical industries, would be a pleasant place to live in or could be turned into a tourist attraction.
>
> (Press release, August 1972.)

Following the rejection of many of these proposals at an oral hearing and given the intensity of this kind of opposition to development from a range of quarters, it is no surprize that most of the proposals failed to materialize, resulting in the stagnation of much of this part of docklands right through the 1970s.

Despite claims in the early 1970s that such development would run contrary to stated government policy aimed at regional dispersal of economic activity, by the end of the decade the global and national economic climate had changed so radically that the potential of this district was again appearing on the government radar. In 1979, Senator Noel Mulcahy presented proposals to an interdepartmental inner city committee established by central government to consider a future for Dublin, within which the Custom House Docks area featured prominently. His report summarized the overall needs of the inner city and argued that the two unused docks at the Custom House could be turned into a fishing harbour by opening them up to trawlers, and this could potentially be accompanied by a village-style development on the rest of the site. The document was sent to Martin Mansergh, a senior government adviser responsible for drafting a discussion report on 'The Inner City Problem'. No immediate policy responses were forthcoming, yet it is highly probable that this report became crucial to the Haughey administration a number of years later in negotiations with the local public representative, Tony Gregory.

The poor reception and negative response with which the 1972 proposals by the Port and Docks Board had been greeted did not deter this agency from trying again to maximize the potential return from the disused port lands and by 1980, a new proposal had been drafted and submitted to Dublin Corporation, the planning authority, for consideration. The plan commissioned from Scott Tallon Walker, the well-known architectural firm that had already played a significant role in shaping the city through buildings such as Donnybrook Bus Garage, Inchicore Chassis Works and Busáras (central bus station), was radically different from its predecessor. It focused specifically on maximizing the development potential of the Custom House Docks, the most westerly part of the modern port, equivalent in size to St Stephen's Green, and a building complex that was at this time over 1.3km away from the rest of the port lands north of the river. These docks, which in their last incarnation had been used for storage by the coal trade, had been abandoned in 1973 and the earlier port proposals had been designed to anticipate this event. The overall tenet of the new plan was that this district could be redeveloped with city centre activities to support the continuing development of the Central

43 The Irish Life Centre, Abbey Street under construction, 1977. (A.J. Parker.)

44 The docklands delimited by the construction of the Talbot Memorial bridge, 1980. (J. Brady.)

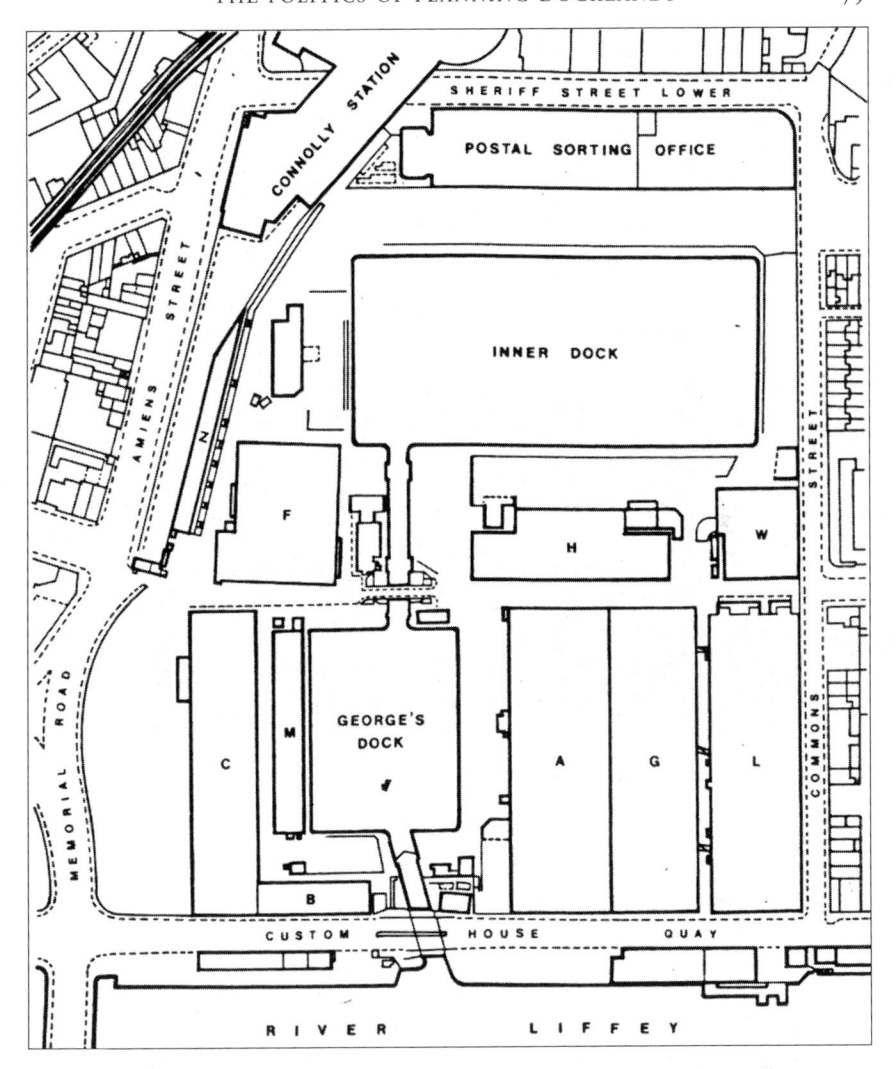

45 Structure of the Custom House Docks site. (Scott Tallon Walker, 1980.)

Business District (CBD) eastwards (Figure 43). The construction of the central bus station (1953), Liberty Hall (1965) the Irish Life Centre (1979) and the continued, speculative acquisition of properties by Irish Life Assurance on George's Quay, south of the Custom House Docks, were cited as key examples of the way in which new trends in retail and commercial development were beginning to emerge in Dublin. The Port and Docks Board scheme appeared to be an ideal project to continue this eastern migration of the city that had begun in earnest from the late-eighteenth century.

Thus the planning application submitted to Dublin Corporation proposed a mixed-use development, with the highest building to be of nine storeys. In the context of its near neighbour, Liberty Hall, this appeared relatively conservative and an appropriate scale to complement the adjacent Custom House and Busáras. Part of the redevelopment would involve the removal of the ramp at Connolly Station, renovation of the main station concourse and the provision of two levels of shopping facilities. A dual carriageway would run through the centre of the site facilitating the easy movement of traffic, while both docks would be used as underground parking facilities catering particularly for rail passengers. South of the dual carriageway at George's Dock, a hotel was proposed to face onto Stack A warehouse which would function as a trade exhibition hall. Additionally, offices and 'living-over-the-shop' buildings would complete development on the southern part of the site with the building height stepped down to five storeys closer to the river.

Reinstating the role of the waterfront as a key focal point within the development would be crucial to success and thus a suggestion was made to construct a floating marina that would accommodate 76 boats, as well as improving and softening the environs through a tree-planting programme along the quayside. The development theoretically conformed to a mixed-use typology, but offices and commercial space were the most dominant activities planned for the site, no doubt influenced by the new developments that had already been completed adjacent to this site but also perhaps an early attempt to be responsive to the changing economic requirements of the city (Table 1).

Table 1 Proposed mixed-use plan for the Custom House Docks, 1980.

Land use	Area (sq. ft.)	Area (sq. m.)
Offices/commercial	2,000,000	185,806
Industrial	450,000	41,806
Shopping	250,000	23,226
Housing	150,000	13,935
Stores	150,000	13,935
Total	3,000,000	278,708

(Dublin Port and Docks Board, 1980.)

The controversy that this proposal generated in 1980/81 now seems odd considering the type and scale of development that has since been undertaken

46 New office block on Mespil Road, 1982. (J. Brady.)

47 Need for renewal on St Stephen's Green, 1983. (J Brady.)

48 Parnell Street, site of the former Williams and Woods factory, mid 1980s. (J. Brady.)

in the capital and indeed in other towns and cities across Ireland. At a meeting of the City Council on 1 September 1980 the vocal city councillor, Sean D. Dublin-Bay Rockall Loftus, became extremely exercized on the appropriateness of this proposal. He questioned (perhaps naively given what we now know about land deals undertaken in the city at the time and the way in which the property market has since operated) whether the Dublin Port and Docks Board had the right to obtain planning permission for such a development with the purpose of then selling the land at a higher value. In a response made at a subsequent meeting the City Manager, Frank Feely, reported that the Dublin Port and Docks Board were within their legal right to sell off surplus lands and he argued further that the 'Port and Docks Board have no obligation to develop but it is in the interest of the Board and the city that development be planned in outline at this stage' (Minutes of Dublin City Council, 3 November 1980).

While the port authority acknowledged that their plan was not in accordance with the existing zoning prescribed in the development plan, they argued that such zoning had become redundant in a rapidly changing economy. A cogent argument for changing the zoning had been made and a liaison officer had been appointed by Dublin Corporation to work with the Dublin Port and Docks Board on their application, nonetheless the request for outline planning permission was denied. It might be argued that this was politically influenced by the vehement opposition from the elected councillors to the proposal, but with it a potential investment of £200 million in the inner city was forfeited at a time when investment was increasingly difficult to generate. In defence of the decision and highlighting the clearly political nature of the decision, the local authority justified it by stating that 'the proposals submitted for outline planning permission were not fully acceptable to the Corporation Planning Officers, or to the members [i.e. the councillors], whose approval was required as the proposal was not in accordance with the zoning in the development plan' (Minutes of Dublin City Council, 6 April 1981). On appeal to An Bord Pleanála, the decision of the Corporation was upheld sounding the death-knell of the first waterfront development proposal in Dublin and paving the way for significantly more controversy, with much greater ramifications, less than one year later.

The need for regeneration

Perhaps the key question to consider is why such a fundamental change in use was considered appropriate in the late 1970s and early 1980s by the Dublin Port and Docks Board? Why did they consider it necessary to move away from their core maritime function and begin to engage in property development and, to some extent, speculation? The key answer is that just as new towns and government policies favouring decentralization had emerged during the 1960s boom, urban decay characterized the recessionary 1970s and early-1980s. Out-migration of industry in the 1970s to new locations on the urban fringe, such as Clondalkin and Tallaght, created an economic vacuum in the city centre. Simultaneously, many new companies establishing a presence in Dublin chose to locate their offices in attractive suburbs such as Ballsbridge rather than the decaying urban core.

The unprecedented rise in unemployment of 97,000 between 1971 and 1981 in Dublin, partially caused by global recession following successive 'oil crises', affected inner urban areas particularly badly. These trends combined with the introduction of new technologies that no longer needed large, extensive and oftentimes polluted sites produced a rapidly changing spatial distribution of economic functions and a redundant inner-city. It appears that Dublin, and other Irish urban areas, were so adversely affected by this shift because of the fragility and character of indigenous industry and the dependence of the economy on specific kinds of foreign enterprise (Telesis, 1982).

This wider economic context meant that, in the early 1980s, local authorities could not afford to undertake significant renewal programmes on their own, nor did the private sector have enough confidence in the market to invest without some form of protection. That is why it is so surprising that the city council at the time adopted such a negative response to the initial Port and Docks Board proposal outlined above, which had the potential to generate new economic activities, rid the city centre of an under-utilized zone and deliver some new housing in an attempt to stem the outflow of population from the core. Nonetheless, the Board had been extremely successful in raising the profile of this part of the city and, to their detriment, highlighting the strategic potential of the Custom House Docks site. What no-one could imagine then is that within a very short time, the future of the site and the surrounding area would lie at the heart of post-election political bargaining in an attempt to form a national government.

The Gregory Deal

When historians and political scientists write about Ireland in the late-twentieth century, it is undoubted that the period 1981–1982 is remembered as one of if not the most politically unstable periods in post-independence Irish political history. Three general elections were held in the Republic in an eighteen-month period because both of the large political parties found it impossible to sustain a parliamentary majority. In early March 1982, after an election that had produced a hung Dáil, neither Garret FitzGerald nor Charles J. Haughey, the leaders of the main political parties, could achieve an overall majority to form a national government and alternative ways of attaining this goal were considered. Within this context, the power of independent politicians acquired a major significance. No individual became more influential than the newly elected Independent TD for northeast inner-city Dublin, Tony Gregory. A former member of the Dublin Port and Docks Board, and a member of the City Council since 1979, Gregory was elected to office having promised to deliver for his constituents on issues including unemployment, housing and education improvements, the removal of derelict sites and the prevention of speculative profits being made by private developers. Through his enterprising approach to negotiations with the leaders of both major political parties, a struggle to gain national political power suddenly became intimately connected with the fortunes of inner city Dublin and provided a unique context within which urban regeneration was about to become headline news.

49 Media reaction to the 'Gregory Deal'. (*Irish Times*, 10 March 1982.)

Prior to the general election and after the failure of the Port and Docks Board proposal, Garret FitzGerald in his capacity as Taoiseach had begun to consider the potential of the area and initiated an investigation into the possible acquisition and development of the Custom House Docks site, in part for social purposes. He envisaged half the site being allocated to local authority accommodation with a minimum of 400 dwellings to be constructed and, in an attempt to broaden usage of this central city location, a two-acre leisure centre with a pop/concert hall and conference facilities. Thus, after the election FitzGerald, in his bid to woo the support of Tony Gregory, was very quickly able to compile a 50-page document containing dramatic proposals worth over £500,000 in support of inner city development. A core part of his bid was the establishment of two urban development commissions, one in Dublin that would draw on £350,000 of government funding to further its objectives and the other at some other, yet to be determined, location within the country. As an incentive and in a clear attempt to obtain the support of Gregory by enabling him to deliver on his electoral promises, FitzGerald suggested the introduction of a 5 per cent tax on derelict sites, an additional allocation of £15 million to Dublin Corporation to construct 500 dwellings in

1982 and a further 500 in 1983, the establishment of educational task forces for deprived areas and, in a bid to combat high levels of unemployment, a £500,000 incentive scheme for businesses who employed school leavers or long term unemployed residents from the inner city. However, the ability of FitzGerald to deliver on all of these pledges was called into question by Gregory. In interview, Tony Gregory noted that:

> He [FitzGerald] came back and he told us he was going to build so many houses on the 27 acre site and of course we knew the housing per acreage and so on and we knew he couldn't do it. He was just trying to bullshit us with things that couldn't be achieved on that particular site to outbid Haughey or whatever. There were no specifics, no timetables, no anything, all generalities.

Gregory's negotiating team, which had initially been more favourably disposed to the political philosophy of FitzGerald, continued in their attempts to obtain the best deal for his constituents and opened negotiations with the leader of Fianna Fáil, Charles Haughey. Even though the first negotiations had been undertaken with FitzGerald, Gregory himself was more comfortable with the kinds of political philosophy espoused at the time by Haughey, who was perceived by many to be more a man of the people. Whether or not this had any bearing on the eventual outcome is unclear, but Gregory is very clear that he adopted an entirely pragmatic approach to the proceedings, describing the deal as nothing more than:

> a response to the neglect of the constituency … The constituency as a whole wasn't disadvantaged, it was one corner that was. And by and large, I had been elected by the neglected corner, which was a feat in itself because by and large the people there didn't vote.

What became known as the 'Gregory Deal' caused uproar between the political parties, as it was argued that Gregory had effectively held the Government to ransom (Figure 49). In defending the estimated £80 million pledged to finance the agreement, Haughey stated that his motives were not just about gaining power but were in the interests of the whole nation. This theme was perpetuated and expanded by many, including opposition politicians such as O.J. Flanagan, who argued that:

as a result of the united efforts of Deputy Gregory and of all Dublin deputies of all parties – the Lord Mayor has an important part which I am sure he will play in this matter – Dublin will be built up and that instead of huge office blocks, many of them foreign financed and foreign-owned, we will have houses and homes so that Dublin people will be able to enjoy living in their own city.

(Dáil Debates, 9 March 1982.)

Explaining his decision, Deputy Gregory argued that his support for a Fianna Fáil government was strategic and in the interests of the city as a whole. It was also aimed at undoing much of the serious damage that had been done to the infrastructure and social life of the city over the previous decade:

> The issues to which Deputy Haughey committed himself included a major increase in Dublin Corporation's housing programme, which has been a scandal for years, the allocation of £91 million for housing in 1982, and a commitment to reach 2,000 houses by 1984 was given. Four hundred new houses in the north centre city area will be started this year ... An almost total breakdown of Dublin Corporation services will now be averted as a result of a commitment by the Leader of Fianna Fáil to allocate a further £20 million to Dublin Corporation's budget for this year. On the issues of employment we put specific proposals to Deputy Haughey. He committed himself to an immediate work force of 500 men costing £4 million for a corporation environmental works scheme and more than 150 additional craftsmen at a cost of £1,500,000 in addition to the present staff to be employed and to give a major boost to the corporation's repairs and maintenance service ... The controversial and destructive motor way plan will not now be proceeded with. The vital 27 acres on the Port and Docks Board site will be nationalized and developed along lines geared to the needs of centre city communities. In the field of education a major commitment to pre-school education along with the provision of a £3 million community school for the neglected centre city area was given, this being part of the designation of the central city area as an educational priority area. Advances in the taxing of derelict sites, office developments, financial institutions and development land were agreed to. A national community development agency will be set up for a budget of £2

million to replace and continue the work of the Combat Poverty Committee ... Once a Government have been elected they will receive my support only in so far as they pursue the programme of agreed commitments and other acceptable policies to me.

(Dáil Debates, 9 March 1982.)

Understandably, members of the Opposition and other individuals and agencies were furious, and this furore was further stoked by Haughey's announcement in a subsequent parliamentary debate that he would not lodge a copy of the deal in the Dáil library for consultation but would instead report orally on the pledges. He informed the house that 42 specific commitments to additional government expenditure had been made particularly in areas such as improved housing, increased funding of the local authority and educational programmes. Most controversially following the recent history of the site, the Custom House Docks complex would be 'nationalized' and a mix of offices, industry, local authority housing and a multi-purpose leisure centre constructed. Part of the proposal was that the Custom House Docks should be assessed for acquisition at its current use value, in other words how much it was worth as a warehousing complex rather than to its potential value as a mixed-use site. The Dublin Port and Docks Board were enraged, as their plans to obtain outline planning permission for a similar mixed-use development had been earlier stymied. Had they disposed of the land with planning permission in place, they would have earned close to £40 million. Under the new deal, the site would be compulsorily purchased for about a quarter of that value, a decision defended by Gregory as a just outcome: 'the Port and Docks Board don't give a damn for the people ... Their only interest has been to sell that site and make as big a profit as they could' (*Irish Times*, 11 March 1982).

Table 2 Proposed location of new housing in north inner-city, 1982.

Location	No. of houses
Seville Place/Oriel Street	33
Mountjoy Street	31
Portland Row/Empress Place	24
Glasnevin	80
Rutland Street	60
Designated Areas A, B, C, D	213

(*Irish Times*, 10 March 1982.)

He argued that if the original plans had been implemented, the inner city would rapidly have become yuppified or gentrified, as long-term residents would have been forced out by rising land values and the construction of luxury flats. Improved and more extensive local authority housing was his key concern, and the first commitment he wanted realized was the immediate construction of 440 houses in the north inner city, in a range of locations (Table 2). The windfall of £91 million pledged to Dublin Corporation to achieve this objective was greeted with disbelief: 'it is difficult to understand how last week the Corporation was just ticking over but this week is jet propelled' (*Irish Times*, 10 March 1982). The Dublin Port and Docks Board were incensed by the deal but they were not alone. Local authorities and interest groups from the rest of the country were also incredulous. Cork city manager, T.J. McHugh, argued for the principle of equity to be applied because 'our financial problems are similar and the social and economic problems in Dublin, Cork and elsewhere are not dissimilar', while IFA President, Donal Cashman, warned that 'if the Government could make funds available for Mr Gregory's programme, then any suggestions that the cupboard is bare for farming doesn't wash at all' (*Irish Times*, 11 March 1982).

The agreement on housing, while a win in terms of social provision, reinforced yet again the role of the local authority in housing provision. While the city council now had the financial means to begin development throughout the city, it did not change the fact that the private sector still displayed no interest in residential development in the inner city, and particularly in docklands, through the 1970s and first half of the 1980s (Dublin Corporation, 1986). The largest of the schemes was undertaken in the City Quay area in two phases, the first in the late 1970s and the second from 1982 to 1984 (Figure 50), a direct benefit of the extra finance made available through the 'Gregory Deal'. The initial phase constituted just seventy-six new dwellings, a mixture of one, two, three and four bedroom units and twelve senior citizen dwellings. Local authority planning applications illustrate that phase two was more extensive comprising two and two-and-a-half storey dwellings on City Quay, Lombard Court, Lombard Street East, Lombard Street West, Dowling Court, Townsend Street, Princes Street South and Creighton Street (Figure 51). An additional 16 units were constructed on Pearse Street in 1983, another four on Macken Street in 1984, and a Corporation scheme of seven high quality houses and 31 flats for senior citizens – Cambridge Court – was completed in Ringsend in 1985, yet this barely began to scrape the surface of the housing and other social problems within the city (Figure 52).

50 Local authority housing development at City Quay. (N. Moore.)

51 Local authority housing on Townsend Street. (N. Moore.)

52 Cambridge Court, Ringsend. (N. Moore.)

53 Controversial reception given to the Urban Development Areas Bill. (*Irish Times*, 24 June 1982.)

Urban Bill condemned by councillors

By Frank Kilfeather, Local Government Correspondent

THE Urban Development Areas Bill, which proposes to set up two urban commissions in Dublin, was strongly opposed by Dublin city councillors at a special meeting yesterday because they claimed it was "totally undemocratic" and would diminish their responsibility.

Describing the Bill as "political expediency," the members unanimously passed a Labour motion, stressing that the Government move was quite uncalled for. The motion pointed out that the council's inner city committee had shown that it was quite possible to carry out redevelopment work in the inner city successfully if given the proper financial backing from the Government.

The Bill proposes two autonomous commissions. One is to oversee the development of the controversial 27-acre Port and Docks Board site near the Custom House for residential and industrial development. The second is to take charge of the development of an area taking in the old "walled area" of the city, including a large portion of the Liberties.

Mr Tomas Mac Giolla said that the Minister had stated that this would not interfere with the power of the City Council but in the next breath the Minister was saying that the commissions would have the power of a local authority. He believed that this was the first of many other commissions. If this Bill was passed, commissions could be set up in other places throughout the city and the country generally, the councillors claimed.

Other councillors were also strong in their condemnation of the Government's move. Mr Fergus O'Brien said it was just "a white-washing exercise," while Mr Dan Browne claimed that it was another effort to erode their power.

Mr Brendan Lynch said it was an effort to take away their power without going through the proper local government reorganisation which had been talked about for so many years. These proposed commissions would not have any elected representatives on them. The Minister for the Environment could make all the planning decisions.

The councillors have decided to go on a deputation to the Minister to express their annoyance at the Bill.

The key policy outcome of the political wrangling that followed the general election, and for the first time entwined locally based urban issues with the formation and survival of central government, was the Urban Development Areas Bill, 1982. Conscious that government survival depended on the effective implementation of the agreement, appropriate legislation was rapidly drawn up. Amidst widespread criticism and controversy (Figure 53), the Urban Development Areas Bill was brought before the Dáil in June 1982 as a general urban policy document even though some members of the House referred to it as nothing more than the 'canonization of the political deal between Deputy Gregory and Deputy Haughey' (Dáil Debates, 16 June 1982). The emerging tension between those promoting development in the capital, and those who wished to see a more equitable distribution of support around the country, was not helped by the emotional speeches made in support of the Bill during its parliamentary reading:

> Dublin people respect and love their city; build it up for them and make it a place where they can live with their flats, their homes, their schools and other educational facilities and not have it as a huge block of aluminium and glass. Let there be children running around and life in the city and do not have many parts of it used for only six or eight hours a day.
>
> (Deputy O.J. Flanagan, Dáil Debates, 9 March 1982.)

Described by some as 'a comprehensive policy of urban renewal and inner city development' (Charles Haughey, Dáil Debates, 9 March 1982), others considered it more 'a tribute to the Taoiseach's wheeling and dealing capacities that ... convinced people to support him here today' (Deputy O'Leary, Dáil Debates, 9 March 1982). There was general agreement that failure to stem the decay of resources, such as community life, commercial centres and infrastructure, and an increasingly apparent rise in crime and vandalism were of general concern. Yet the way in which the Bill was enacted proved hugely divisive. George Birmingham, a Fine Gael representative for Dublin North Central, acknowledged that:

> There is a consensus stretching around the House that the problems of decay and dereliction in this city and elsewhere are rampant. There is recognition of the social ills that stem from that, poor housing, urban crime and unemployment. There is an equal consensus that the existing

methods of dealing with that problem have to date proved inadequate and we must look elsewhere for solutions. I repeat that we are predisposed to be impressed by the Minister but at this point in time we find the Bill is full of flaws.

(Dáil Debates, 16 June 1982.)

The key element at the root of such vehement opposition, was the proposal to establish two special commissions within Dublin City, the first to oversee renewal at the Custom House Docks site, and the second to promote regeneration in the 'walled' or Liberties area of the city. The second critical issue was the proposal to remove appellate powers in particular designated areas from An Bord Pleanála to the Minister for the Environment of the day, at that time Ray Burke, described by the Port and Docks Board as 'legalized hijacking'. This emerged from the suggestion to exempt development sponsored and approved by the commissions from the traditional planning process, following the model that had been adopted in London and Liverpool docklands in 1981. As pilot projects, the success of the commissions would determine whether similar provisions could be extended to other urban areas. The chief critics of the Bill included Dublin mayor and former senator, Alexis FitzGerald, who robustly opposed it at a City Council debate on 23 June 1982 and An Taisce, the National Trust for Ireland, who foresaw danger in exempting particular sections of the city from the planning process, undermining the role of the local authority and resulting in windfall profits for landowners, the very issue against which Gregory had been elected. The supposed democratic deficit inherent in the operation of these commissions gave rise to an extended debate in Dáil Éireann with various members supporting the rights of local authorities to plan, govern and control the city. The Government countered by arguing that the nationalization of the Custom House Docks site was necessary to safeguard the interests of the citizens as a whole.

In the Dáil debate on 16 June 1982, Deputy Niall Andrews strongly defended the government action. Yet it would appear that in the context within which this Bill was introduced, only a very short time after the Dublin Port and Docks Board had actually attempted to gain planning permission for redevelopment, his arguments were somewhat disingenuous. While Deputy Andrews rightly argued that the Board was 'answerable to people as a semi-State organization, answerable to Dublin people in particular and to the Irish people in general', it appears somewhat ironic that his defence of the government position was built on the argument that 'if, over the years Dublin

54 Georgian Houses on Cumberland Street just prior to demolition, 1980. (J. Brady.)

55 Dublin quays 1987. (J. Brady.)

Port and Docks Board have not developed that site, so be it; the Government must do it. I welcome the fact that the Government plan to do this'. This kind of support was also forthcoming from other quarters, particularly from property developers and auctioneers, who anticipated that this legislation might give a necessary boost to the depressed construction industry. They warmly welcomed the Bill arguing that 'the private sector should be encouraged to participate in the re-building of our inner cities. Special incentives could be offered in order to promote socially useful developments in these areas' (*Irish Times*, 30 June 1982).

However, all of this debate quickly became meaningless as protracted political instability once again undid the proposals that could perhaps have begun to tackle coherently the development of the capital city. With the collapse of the Fianna Fáil government in November 1982, only nine months after the deal had been struck, the vital opportunity that had been dangled in front of residents in the northeast inner city in the early 1980s had been lost. Nonetheless, Tony Gregory still argues that:

> by and large whatever could have been done in that time was done. Housing projects were given the go-ahead: a lot of new urban renewal housing in Seville Place and the Summerhill area, and not just on the northside but right around the city, was sanctioned. The money was given to Dublin Corporation to start them and once they started there was no stopping them, and there was no stopping them for several years, because they were all on the planning boards, they were ready to go. The Corporation had been starved of funds to develop this new idea of inner city housing. They were given the funds and they got the contracts underway and it flowed from there.

When asked why he believed Charles Haughey ensured that money was released to the Corporation and the Urban Development Areas Bill set in motion so quickly, Gregory gives a very pragmatic, rather than any sort of ideological or welfare, explanation:

> It wasn't done because he was a nice guy or anything. It was done because he knew he had to do it to stay in power because he had been convinced of this by his brother in the city council who warned him that the one thing not to do with me was bullshit me. If he did, I'd pull the plug and change sides overnight. And so he knew it. He knew from

talking to the group that he would be gone, so he didn't. He was just a pragmatist, he said to us, 'You know what I want, so what do you want?' Because we did know what he wanted; he wanted power. To me, my reason for being there was that I was elected to do something about these issues.

Nonetheless with the fall of the Government after only a few months, the 27-acre Custom House Docks project was put on hold, and the site remained derelict for a further five years. The fate of this area was very clearly off the national agenda until the passage of the Urban Renewal Act in 1986. Unlike previous attempts at redevelopment that had been very clearly tied up with local and national political influences, the development of the docks from 1987 was very much a product of international influences as the Irish Government looked to the experience and success of other countries in regenerating similar areas. If the initial application of the Port and Docks Board had received the go-ahead or if the Urban Development Areas Bill had been enacted and implemented, it is likely that Dublin would have been one of the earliest cities to capitalize on the potential of redeveloping the waterfront zone. All of these delays ensured that by the mid to late 1980s when the docklands project was eventually reignited, Dublin was very much following, rather than leading, international trends in urban regeneration and redevelopment.

Urban Renewal Act, 1986

Unless there is a specific policy of urban renewal we are heading for a crime-ridden empty shell in the inner city of Dublin with a decline in population, falling school numbers and schools in bad repair... [this is] a crime against our citizens and not just our urban citizens, because all our taxpayers in the long run will have to pay for the cost in terms of crime and vandalism and in terms of ultimate renewal of infrastructure if we do not come to grips with this problem even at this late stage.
(Deputy Gay Mitchell, Dáil Debates, 10 June 1987.)

Without intervention in planning and the economic development of towns and cities through most of the first half of the 1980s, Irish urban areas had continued a pattern of economic, social and physical stagnation that began in the early 1970s. The scale of the problem was widely recognized. As early as 1975, the Royal Institute of Architects of Ireland had published *Dublin: A City in Crisis* indicating the problems and possible solutions, but almost a decade later, in 1986, by the time a Dublin Crisis Conference was held in the Synod Hall at Christ Church not much had changed. The Conference was attended and addressed by the Taoiseach, Garret FitzGerald, who made a number of pledges to conservationists, planners and others concerned about the future potential and viability of the city centre. This wider context provided to some extent a very welcome environment in which to introduce a new Urban Renewal Bill. The policy proposals were dramatic as they represented a shift in the operation and purpose of planning, and some of the first serious attempts by any Irish Government to be pro-active and pre-empt urban development. Secondly, it represented a total reversal of earlier government policies of decentralization as it promised to stem the outflow of human and financial capital from the inner city and encourage the development of new urban functions in older areas. Perhaps drawing on the experience of the Urban Development Corporations (UDCs) that had been piloted in the United Kingdom and were becoming more common in the United States, Irish renewal programmes aimed to remove as many development constraints as possible to encourage private sector activity. But like the Urban Development Areas Bill introduced in 1982, the Urban Renewal Bill was not enacted without difficulty and much controversy. Central to this were three key factors: firstly, there was widespread disagreement as to the areas that should be designated for tax incentive purposes, a highly politically charged issue; secondly, the management of the designated areas raised some concern; and the third major difficulty was with the type of development that might take place.

As well as specifically providing for a new management and financial regime in the docklands area, three other areas in Dublin were designated for financial incentives – Henrietta Street; the north and south quays stretching from St James' Gate to O'Connell Bridge; and the Custom House Docks site (Figure 56). All areas would benefit from a standard package of tax incentives including capital allowances, double rent deductions and rates remission provided initially in the 1987 Finance Act. The Custom House Docks profited from all of these and a number of additional incentives (Prunty, 1995). In defence of this, the Minister of State, Fergus O'Brien, commented that:

> The Custom House Docks is of significant social and civic importance and could not be left undeveloped without considerable repercussions for the surrounding area ... This site is a jewel in our city. It is a jewel that needs to be treated, polished and developed well. When one

considers that site, with the water element in it, it affords a great opportunity to do something magnificent.

(Seanad Debates, 12 June 1986.)

This potential project had many detractors including those who argued that safeguarding medieval Dublin would have been far more valuable than targeting the docklands site for new development. The establishment of a Dublin Walled City Development Authority, similar to that proposed in the 1982 Bill, was suggested and strongly supported by the opposition but the Government disagreed stating that parts of medieval Dublin were included in the southern quays designated area. The remainder of the area was under the control of Dublin Corporation who, it was argued, were the most appropriate body to look after the historic core. This statement was ridiculed in the Dáil, given the fiasco that had ensued at Wood Quay less than a decade earlier, when the new Civic Offices were constructed over the remains of the medieval centre, but the Government stood firm on their opinion.

The new development agency for the Custom House Docks would oversee redevelopment in partnership with private interests, but their job was made difficult by the resistance to the new institution and the adverse political and media opinion at the time. The potential of docklands as a key site for redevelopment was difficult to envisage and was ridiculed even by those within the governing parties. Liam Skelly, TD, who represented Dublin West and was a member of the governing Fine Gael–Labour Coalition, noted that:

> Growth in the city is towards the west, towards Heuston station. No-one in their right minds would invest in the docks area. The place to invest is in the other designated area along the quays ... Long term investment in the docks area is questionable. Within hours of the docks area being designated, the top property people gave it the thumbs down. Developers are not anxious to spend money there because it is not where the action is.
>
> (Dáil Debates, 28 May 1986.)

Leaving the economics of the decision aside, project management became an even more divisive factor. The Urban Renewal Act proposed that all planning powers within the area were to be transferred from Dublin Corporation to the new Custom House Docks Development Authority, which would report directly to the Minister for the Environment. These close

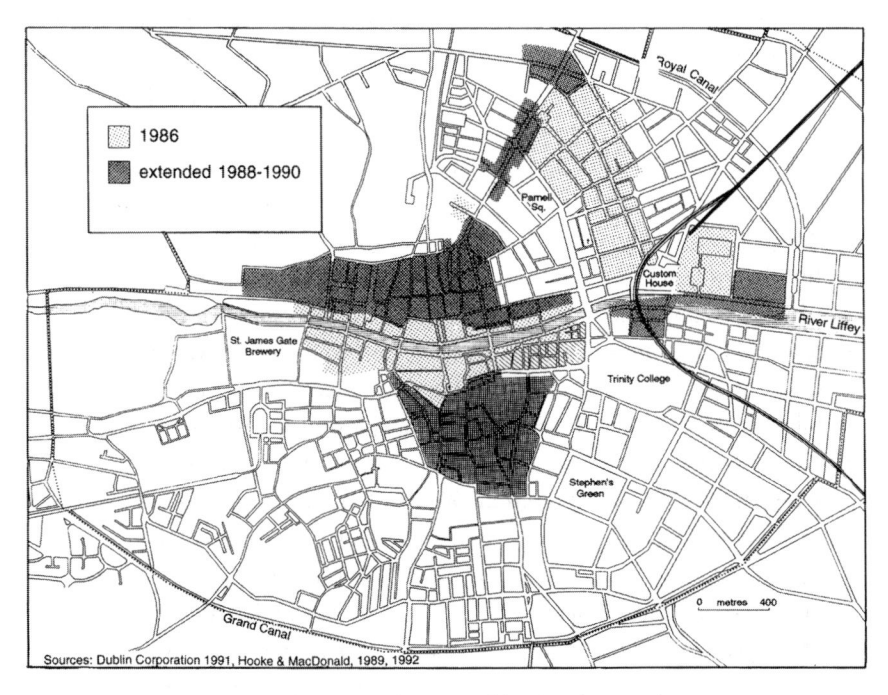

56 Urban renewal areas, Dublin. (Prunty, 1995.)

57 Advertisement for urban development site, inner city Dublin, 1990. (J. Brady.)

links to be established between the authority and the Minister, while bypassing the traditional planning and consultative process, caused great concern. While arguably overly dramatic, the remarks of Tomás MacGiolla, a member of the Workers Party who also represented Dublin West, reflected the general mood of many members of parliament during the debates:

> Dictatorships are in many cases more efficient; but they are doing what the dictator wants, not what the people want. That is precisely what is happening in this Bill. It is concerned with what the Minister wants done, not about the wishes of the people of Dublin. The beginning of this is to wipe out the working class people from the inner city and put in a nouveau riche type with their marinas and boats, following the line of what has been done in London. That is the purpose of this and it worries me exceedingly. The headquarters of the Garda Síochána are to be in the same area to ensure that these people are protected from the marauders on the outside.
>
> (Dáil Debates, 28 May 1986.)

Concerns were also raised regarding the type of development that would occur as the Government seemed to be in favour of mixed-use and commercial development. This was odd as the site was supposedly nationalized in 1982 to prevent exactly this kind of development. In direct contrast to his party's earlier stance which had instigated nationalization for social purposes, Minister Pádraig Flynn in an address to the Seanad on 1 July 1987, described the new legislation as a way to 'create the right environment to lead and encourage development and investment, particularly by the private sector, through utilizing appropriate tax incentives'. His key idea was to ensure a mix of housing tenure, commercial development, boutique shopping and other amenities. These were indeed interesting and impressive aspirations, but it appeared that the aims of the Fianna Fáil party had radically altered in five years. In his response to government proposals, Proinsias de Rossa of the Workers' Party, noted the vagueness of many of the government proposals and remarked that:

> It is clear to me that what is intended is that Sheriff Street will be cleared not only of tenants but of flats. What I am arguing is that we must ensure that those people will be able to live and stay in the area in which they were born and reared. It is my contention that ... we will

have a development on that site which is totally geared towards commercial and up-market apartment living. Experience in other countries has indicated that that is so. The Minister said he intends to see that there will be balanced development on the site. He goes on to say he is not too sure what balanced development means.

(Dáil Debates, 28 May 1986.)

Nonetheless, the landmark proposals were eventually enacted, with perhaps one of the most significant impacts being the unique planning framework that it established for the docklands area.

The Custom House Docks: a unique planning environment

It is difficult to establish exactly the genesis of the Custom House Docks project but it is clear that the broad framework established for wider urban renewal in Dublin was intended to boost the flagging construction industry. It is likely that this objective together with the fact that the Custom House Docks site had been on the government radar since 1982 and international experience that demonstrated the economic potential of waterfront renewal schemes, all combined to position this site favourably in the minds of legislators. The Urban Renewal Act, 1986 provided for broad renewal throughout the city, while singling out the Custom House Docks as a special case. Very clear boundaries were drawn around the area that would be designated for financial incentives additional to those available more generally. The original boundaries ran from the Custom House, east along Custom House Quay to Commons Street; north along Commons Street to the junction with Sheriff Street Lower; west to Amiens Street; and south to Memorial Road. A subsequent amendment in 1987 extended the eastern boundary to Guild Street, what is probably more familiar today as the western edge of the Spencer Dock site, and included the northern half of the River Liffey. This latter inclusion was unusual and did nothing to contribute to the evolution of an overall vision for the riverfront (Moore, 1999). The reasoning appears to have been both economic and political, as Minister Flynn argued that including the entire river in the designation might impede development at George's Quay; how this might occur was never explained. In 1994, another extension was made to the Custom House Docks Area to include the Sheriff Street flats site that had formerly been earmarked for the National Sports Centre. This amendment also empowered the Minister to order extensions of

58 USS *Constellation* and the inner harbor Baltimore. (U. Blosfeld.)

59 Canary Wharf development, London. (W. Gibb.)

the Custom House Docks Area as far as the Point Depot and Alexandra Road, resulting in the fact that in less than a decade, the area under the remit of the Custom House Docks Development Authority was expanded from eleven to thirty hectares (Figure 65, next chapter).

In the parliamentary debates that surrounded the designation of this area, one of the key concerns, as indicated previously, was how development would be managed. At the same time across North America and Britain, schemes like that at the Custom House Docks were being established in many city-centres such as at Baltimore Harbor (Figure 58) or Canary Wharf in London (Figure 59). As well as promoting new physical and economic landscapes, all of these projects produced new regulatory environments. Rather than the traditional authorities, like city or district councils, managing the development, new institutions were established instead. In choosing the best way to manage regeneration in Dublin, the Government followed this trend by establishing the new agency – the Custom House Docks Development Authority (CHDDA) – to oversee development.

This authority was established with one key objective: to rejuvenate this historically significant, yet decaying part of the urban core. Through the implementation of targeted strategies similar to those adopted by the London Docklands Development Corporation, it was anticipated that long-term economic, physical and social problems would be eradicated. The authorities hoped that through the encouragement of local businesses, employment opportunities within the city-region would become diversified and local skills would be mobilized (Benson, 1988). The actual establishment of the authority did nothing to stop the criticism that had surrounded the institution since it was first proposed. In a critique of Irish planning, Bannon (1989, p. 170) has questioned the wisdom of using such a heavily streamlined development corporation as it 'raises important questions as to the nature of the planning approach being used, the relationship to the surrounding local authority, Dublin Corporation, and most of all the relationship to the adjacent communities and their participation in or benefit from such an undertaking'. This emphasis on participation and engagement with local residents became an increasingly controversial bone of contention during the early years of redevelopment. Others also criticized this approach, particularly those with land interests in the area. In order to facilitate the kinds of approach and development being envisaged for the extended Custom House Docks area, control over large portions of land was transferred statutorily from other agencies to the CHDDA (Table 3). In many ways, this was perhaps necessitated by the desire to adopt a comprehensive and integrated approach to development but it did result in even further erosion of the legitimacy of the new agency.

Table 3 Sequence of land transfers to the CHDDA.

Year	Land transferred from
1987	Dublin Port and Docks Board
1988	B&I Shipping Line
1992	Dublin Corporation
1994	Minister of Finance
1994	Dublin Port and Docks Board

(Statutory Instruments, various years.)

In line with the practice in other cities, the CHDDA were charged with the production of an overall strategy or Master Plan, within which the private sector could develop and present their proposals. In June 1987, the authority submitted their first planning scheme to the Minister for the Environment, addressing three key issues – the nature and extent of development, the distribution and location of uses and the overall design of the area including maximum height and external finishes. Lest it be considered that the CHDDA had free reign within the designated areas, they were constrained legislatively. Although they were responsible for the day-to-day running of the project, it was central government who played perhaps the most important role in the area through the sponsorship of financial incentives, the appointment of members of the authority and the approval of planning schemes.

Just as earlier debates on the wisdom of organizing redevelopment in this manner were dismissed, the power of Dublin Corporation was also eradicated from this area. In the United Kingdom in the 1980s, similar arrangements were forming part of successive conservative governments' attempts to 'roll back the State' and diminish the power of local authorities, leading to what has been described as the death of planning. In Ireland, this attempt was met with widespread criticism and concern, as the role of the local authority was with one stroke of a pen reduced to consultation. In drawing up planning schemes, the authority was simply required to 'have regard' for the provisions of Dublin Corporation's Development Plan and in contrast to the general provisions for public participation in Irish planning, legislation governing activity in the Custom House Docks Area provided local residents with little opportunity to influence development. This was particularly ironic for, while espousing the need for this new agency for docklands, the same government disbanded the newly established Dublin Metropolitan Streets Commission. While there were very cogent, and indeed reasonable, reasons for doing this, the arguments used to justify the dissolution of the Commission could just as equally have been levelled at the Custom House Docks Development Authority. In support of the stance of his party leader during the Seanad Debates, Seán Haughey explained that the Government believed that the Dublin Metropolitan Streets Commission should be disbanded as such an agency would be 'an attack on local democracy. The Act establishing the commission provides for consultation with Dublin Corporation, the Dublin Transport Authority and other interested bodies. Mere consultation, however, is not enough' (Seanad Debates, 1 July 1987).

In an attempt to counter some of these concerns, the CHDDA stated a clear intention in the 1987 Planning scheme to establish a consultative community liaison programme during the development phase and to provide financial assistance to local community groups. Yet, the community liaison committee (CLC), comprising members of the authority and representatives of community groups, was not established until 1995, eight years after the project was initiated. The role of the CLC in maximizing the involvement of the communities in redevelopment; providing a forum for direct communications between the authority and the developers with representatives of the local communities; and exploiting suitable employment opportunities for people from the local area appears to have been almost entirely aspirational. This is not unique and is directly in line with the weakness of social provisions in other urban waterfront regeneration schemes around the world. In most of these, consultation with locals has not been ranked highly on the agenda of development corporations, as consultation and participation has traditionally been perceived as one of the key impediments to efficient and timely development, the very problem these agencies were established to rectify.

Internationalising the city: the vision for the Custom House Docks

The attraction of private investment into the docklands area, through the promotion in particular of the development of an International Financial Services Centre, ensured that not only would this area have important symbolic value as a flagship project of urban regeneration in the capital, but it would also have a very significant economic dimension. Nonetheless, this was not unique to Ireland as a plethora of projects have been promoted in the last twenty years in waterfront cities as far apart as South America, Australia, South Africa, the Far East and other parts of Asia (Table 4).

Table 4 Major global waterfront redevelopment projects.

Date	Project/City	Size (ha) approx	Cost (US$ unless other stated)
u.c.	Teleport City, Tokyo	448	
1963	Inner Harbour Baltimore	38	2.5 billion
1972	Harbourfront, Toronto	36	340 million
1979	Granville Island, Vancouver	17	70 million
1979	Battery Park City, New York	37	4 billion
1981	Docklands, London	2024	
1983	Minato mirai 21, Yokohama	186	$200 billion
1988	Darling Harbour, Sydney	60	$2.5 billion
1989	Victoria & Alfred, Cape Town	82	R. 2.5 billion
1989	OJ Havengebied, Amsterdam		$2.5 billion
1990s	Port Vell, Barcelona	54	340 million
1990	Salford Quays, Manchester	60	$750 million
1990s	Kop van Zuid, Rotterdam	125	DFl.475 million

(A. Breen, & D. Rigby, 1996, p. 25.)

Just as they have embraced similar management structures in terms of the establishment of separate development agencies, so too has the type of development promoted in these places been very similar, influenced by the 'festival marketplace' imagery developed in North America.

In Baltimore (Maryland), sometimes referred to as the cradle of waterfront regeneration, the Inner Harbor has been regenerated with a combination of specialist retailing, sporting activities, offices and new residential areas. Formerly one of the most dangerous and undesirable parts of the downtown area with a reputation for drugs, prostitution and violence, it has been reborn as a chic location for well-to-do residents, employees and daytrippers. At South Street Seaport, New York close to the former World Trade Centre, a mix of museums, shopping, restaurants and new loft apartments has been promoted, while at Cardiff Bay in Wales, the new Welsh National Assembly Building stands close to Techniquest (an interactive science museum), restaurants, shops and waterfront apartments (Figure 60).

60 Techniquest science museum, Cardiff. (N. Moore.)

61 Artist's impression of docklands redevelopment. (CHDDA.)

This emphasis on a wide range of land uses throughout all of the projects was quite deliberate and designed to attract a range of users throughout the day and night. In addition, many of the projects were also designed by the same architectural practice, Benjamin Thompson and Associates. Based in Boston, Thompson had worked through the 1960s on developing what he termed, 'The City of Man'. His argument was that all cities should be people-places, designed at a human-scale with an emphasis on social interaction, and an awareness of nature: 'of changing seasons, of orientation to water, and of places intimate by day and radiant by night, where the lovely unpredictability of life would be nurtured and experienced on a daily basis' (*Boston Globe*, 19 August 2002). Since the ultimate objective of the Custom House Docks Development Authority was the development of a high quality, vibrant urban environment, Thompson was considered the ideal urban designer to produce the Master Plan. His vision took account of the existing morphology of the area, to 'blend new buildings with Dublin's rich historic heritage' and move 'beyond integrated components for living, working, shopping, culture and recreation ... to create new civic spaces and reclaim major frontages on the river Liffey for public access, day and night' (BTA Architects, Portfolio 2003).

Re-defining the relationship between the river and the city and re-integrating the Liffey as an essential element in the urban environment was central to the Dublin project. In terms of its overall design, Trinity College was suggested as a potential model for successful urban living as it was substantially pedestrianised, entered from well-defined gateways, and operated successfully as both a public cultural and private functional space. In line with Thompson's vision of the city as a place for human interaction, the pedestrianization of, and re-routing of traffic along, Custom House Quay was suggested as a medium-term objective because 'a pedestrianized environment would consolidate the basis for an extended riverfront environment in Dublin's city centre' (CHDDA, 1987, p. 5). The elimination of dead ends, and the construction of a pedestrian underpass to improve safety in Beresford Place were considered essential to the creation of a high quality urban environment (Figure 61), 'which [would form] part of a broader vision in which the city asserts its qualities as a place in which to work and live' (CHDDA, 1987, p. 15).

In order to ensure a range of uses were developed at the Custom House Docks site, office/commercial, residential, retail and cultural/amenity functions were zoned in the Master Plan. The core commercial element would be an International Financial Services Centre (IFSC), located on the southwest corner of the site closest to Gandon's Custom House. As a 'flagship project'

62 This office block on Mountjoy Square seemed destined never to be completed, 1990.
(J. Brady.)

the IFSC was expected to encourage the vibrancy and viability of the entire development by acting as a magnet for investors, a relatively risky decision at a time when the Dublin office market was suffering from over-supply (Figure 62). Aware of the large number of small apartments proposed for other designated areas and wanting to differentiate docklands from other renewal areas, the CHDDA aspired to the provision of a range of types and sizes of accommodation, including suitable residences for families. The proposed location was the Inner Dock, away from the main traffic arteries and a suitable location in which to exploit the amenity value of the water body, a factor that had proven significant in attracting residents to other docklands locations, such as London.

To ensure that the area would become an important attraction for people from other parts of the city, the authority strongly encouraged the introduction of specialist retailing to the area. The development of a 'festival marketplace', as in Tempozan, Japan and Baltimore's Inner Harbour, would contribute to the diversity and vibrancy of the area and was a type of development favoured strongly by the architect. Food halls and small craft shops located in covered concourses and atria (similar to the Winter Garden at Battery Park City or Quincy Market in Boston) aimed to attract highly-paid professionals and tourists to the area.

Table 5 Major land uses in the Custom House Docks.

Activity	Area (sq. metres)
Financial services	28,427
Commercial offices	42,180
Retail	12,450
Residential	18,950
Cultural	12,170
Hotel/Conference	27,870
Car parking spaces	(1,813)
Total scheme (excl. car parking)	142,047

(Adapted from Bannon, 1989, p. 168.)

63 Aerial view of docklands, 1986. (DDDA.)

64 Derelict interior of Stack A warehouse. (N. Moore.)

The final and critical element that would contribute to vitality and vibrancy in the area at all times of the day and into the night hours would be cultural rejuvenation. Stack A, a historic and architecturally important nineteenth-century warehouse, was considered the key component in delivering this element of the scheme even though it had fallen into a state of disrepair (Figure 64). The restoration of this structure and its unique character would help to attract a critical mass of visitors to ensure the viability of other amenities such as water-based activities, restaurants and specialist retail facilities. Another suggestion permanently to moor a historic vessel in George's Dock as an informal museum closely followed the North American examples of New York and Baltimore, and St Katharine's Dock in London, where historic ships are moored year-round. All in all, a broad-based strategy of construction, renewal and conservation was envisaged for the Custom House Docks area that would certainly have economic benefits at a city and national, if not a local, scale.

However, given the organic and dynamic nature of cities as well as political indecision regarding the appropriate extent of the designated area, the plans were overtaken by reality. Even as the original Master Plan was being drawn up, the Government had changed the boundaries in an amendment to the

Urban Renewal Act and as such, the 1987 Master Plan was almost out-of-date before it had even been completed. Nonetheless, the foundations laid through all of the political wranglings of the 1980s, the making operational of the Urban Renewal Act, 1986, the establishment of a new planning environment for the Custom House Docks and the subsequent publication of the Master Plan, were to have a tremendous influence on development in the immediate area and its hinterland over the next decade. The influence of the development that occurred through the 1990s is still being felt within the docklands zone and has been considered a major part of the Celtic Tiger economic boom, as well as one of the driving forces of current development patterns in this area.

Recreating the waterfront: the Custom House Docks and environs, 1987–1997

> In ten years time I'd like to see it as being the start of the whole redevelopment of the docklands area of Dublin and the inclusion of that area in the city itself, because this particular site of 21 acres is less than half a mile from O'Connell Street. I really think it's the start of something much bigger.
> (Mark Kavanagh, chairman of the CHDDC on *Face of the Earth*, RTÉ television, 1 March 1988.)

The optimism displayed by Mark Kavanagh, chairman of the company that won the tender for the Custom House Docks redevelopment signalled a major change in perception over the potential and future development of this part of inner city Dublin. As is clear from the previous chapter, until the designation in the Urban Renewal Act of the Custom House Docks for redevelopment, little private sector interest in redeveloping the docklands had been apparent; that began to change gradually as the private sector became increasingly aware of the development potential of the site, the newfound significance of adjacent areas and the availability of financial incentives, provided by the State, for investors. Local representatives such as Father Frank Duggan, Chair of the North Wall Community Association, concurred with those closely involved in the delivery of the redevelopment process that this would be the start of a much more extensive development programme. He explained that:

> The fact that our people are denied the opportunity for jobs in reality, the fact that the docks labour force has dwindled dramatically over the years; all of these things have made of this area a ghetto. The people see in it [the redevelopment] the chance to get out of it. I am very optimistic because the people here in the last six months have shown a remarkable change in their own attitudes, a sense of hope is creeping in where there was none ... a sense of a better future for the whole

community here, which would benefit the whole city of Dublin and the whole country.

(*Face of the Earth*, RTÉ television, 1 March 1988.)

Others, while generally positive, sounded a note of caution. Mick Rafferty from the Alliance for Work Forum, a long-term associate of Tony Gregory and one of the key negotiators of the Gregory Deal that had been agreed in 1982 with Charles Haughey, remained cautious. Laying out his vision for the future of the Custom House Docks Area he hoped that:

> In five or ten years time, I'd like to see all the little joyriders down in Sheriff Street going around carrying filofaxes … The short-term benefits are that construction jobs will be created on the site and FÁS have agreed through negotiations with the unions and the Custom House Authority that unqualified school leavers, for instance, will get tested for apprenticeships. That's a major step. Secondly there has been agreement with the Custom House Authority and therefore the developers … that preference will be given to local people assuming that they have the skills so there is a need for FÁS to have induction courses … Long-term benefits are the real problem. Hopefully they will recognize that the site cannot develop without the community that exist alongside it developing. The long-term benefits should be an increase in the housing quality of people plus securing some of the long-term jobs. But that is a huge social engineering job that has to be done and there is no sign that the Custom House Authority are applying the same energy, skills and whatever to solving some of these social problems as they have to attracting development into the site in the first place.
>
> (*Face of the Earth*, RTÉ television, 1 March 1988.)

The note of caution sounded by such local activists was quite contrary to the reception given to the proposals by the property industry. Speculators and other business interests were very positive about the potential of the new plan and indicated a determination to maximize the development opportunities at this landmark site. One of the key drivers of a plan to establish a new commercial heart at the Custom House Docks, as part of an overall redevelopment strategy, was the prominent businessman Dermot Desmond. His stature, while considerable at the time, has grown tremendously since. In 2006, *Forbes* magazine estimated his worth at €1.419 billion; he is ranked fifth on the Irish Rich List and is listed number 746 among 'The World's Richest People'. He had founded NCB Stockbrokers in 1981, Ireland's largest independent stockbroker until it was sold in 1994 to the National Westminster Bank, now the Royal Bank of Scotland. This experience meant that he was uniquely placed to identify the needs of the financial industry in Ireland in the 1980s and the conditions necessary to encourage growth. Widely considered to have been the first promoter of the idea of a new, designated financial heart for the city, his access to the corridors of power in the early 1980s ensured that this idea was quickly taken on board. The idea of an International Financial Services Centre became not just part of the overall plan for the docklands but the major flagship project. What is surprising, given the lack of speed with which the organs of the State usually grind, is that within a very short time the idea of an International Financial Services Centre became a keystone of the entire regeneration project and today ranks as a key symbol of the 'new' docklands.

Of most surprise in the 1980s, was the rapidity and success with which direct government intervention in this area created a huge momentum. The welcome afforded to the Custom House Docks project was such that initial plans rapidly became obsolete. As previously noted, even before the original Master Plan was drawn up in 1987, the Government gave the first clear signal that they had reason to be hugely confident in this venture by ordering an extension to the boundaries established in the Urban Renewal Act. Given this, the 1987 Master Plan was almost out of date before it had even been completed and a new planning scheme was published in November 1994 encouraging high-density, commercial development.

The Sheriff Street Flats, one of the largest local authority housing schemes within the city where major social problems were rampant, was adjacent to the original Custom House Docks site but had been omitted in the original planning scheme. In addition, the National Sports Centre site, the An Post central sorting office, Connolly Station and the northern half of the Liffey along Custom House Quay had also been omitted but all of these were included in the second plan (Figure 65). Yet even though the geographical extent of the project area was extended, the scope of the project appeared to have contracted by 1994. While the 1987 scheme aspired to economic, physical and social rejuvenation, the primary emphasis of the second scheme was on economic regeneration, specifically the facilitation of continued expansion at the Financial Services Centre. Even in discussing the non-commercial elements of the plan, it is clear from the wording of the 1994 Planning Scheme that the

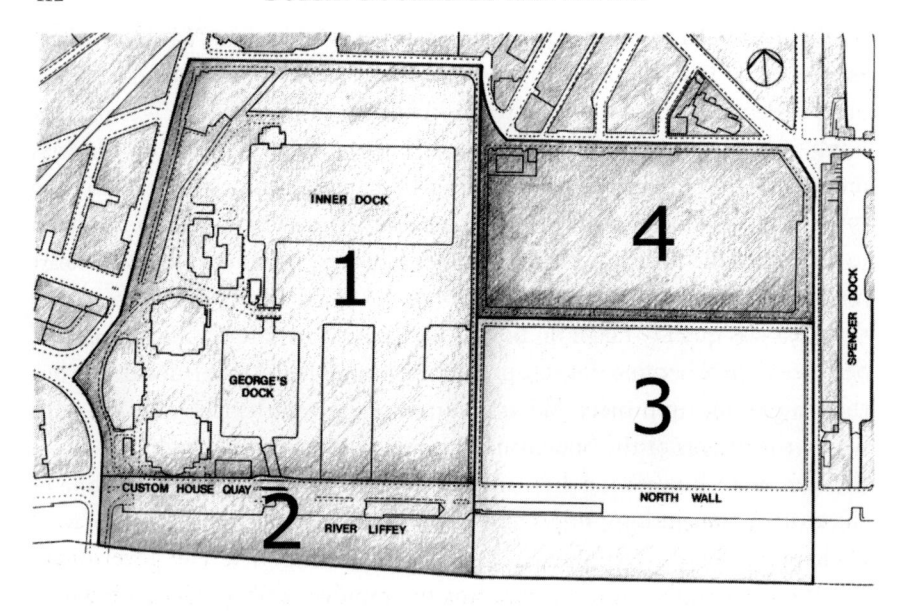

65 Process of extension of Custom House Docks area. (CHDDA, 1994.)
Key: 1. Custom House Docks area (Urban Renewal Act, 1986); **2.** Custom House Docks area first extension (1987); **3.** Custom House Docks area second extension (1988); **4.** Custom House Docks area third extension (Urban Renewal (Amendment) Act, 1987).

winning proposals for development would be those considered most compatible with the continued development of the IFSC. So, how did this one element grow to have such a pivotal role in the overall shaping of the docklands regeneration?

Genesis of the International Financial Services Centre

Mirroring similar developments in other waterfront locations, and in particular that of Canary Wharf in London docklands, the International Financial Services Centre project was designed to attract international investment to Dublin with the overall aim of improving tax revenue and jump-starting an ailing national economy. The Taoiseach of the day, Charles Haughey, described it as 'a centre with the technology, communications and right regulatory background [which] will allow us to carve out our own little niche in this expanding market ... and make an exceptional contribution to growth, investment and employment' (*The Times*, London, 19 April 1987). Most interestingly, although Charles Haughey is credited with the vision for delivering on this project soon after his election in Spring 1987, it was the previous Fine Gael–Labour coalition government that had initially explored the opportunities for a similar development and undertaken some of the groundwork. In 1986, Minister for Labour, Ruairí Quinn, and Dermot Desmond of NCB Stockbrokers jointly funded a study by Price Waterhouse to investigate the feasibility of a financial centre in the city, but the conclusion was that such a project was doomed to failure. Nonetheless, it seems that Desmond was determined that such a project had potential, and prior to the general election of 1987 he brought the idea to Charles Haughey. As Leader of the Opposition and a TD for North Dublin, Haughey was interested in exploring any options for reversing the seemingly interminable problems with the national economy that had generated significant societal difficulties including high unemployment and other social problems. The idea of a financial services centre was a keystone of his pre-election Programme for Government and was one of the first elements of the Programme developed by the new Fianna Fáil government following general election success. In a recent news article, Michael Buckley, formerly chief executive of AIB Bank and a close associate of Dermot Desmond at the time, outlined the extent of the personal interest taken in the development of the idea by the new Taoiseach:

> At the beginning of the first meeting, Ó hUiginn [advisor to the Taoiseach] made it clear to us all that 'the Boss', as the personal sponsor of the project, would take a deep interest in how it went, and would be upset indeed if there was any outbreak of silo-ism, unnecessary bureaucracy, or sidelining of good ideas. He made it clear that his job, as chairman of the group, was to make sure our recommendations went straight to the Taoiseach. If he was happy with them, they would be followed with legislation or regulatory action, with unusual expedition.
> (*Sunday Business Post*, 18 June 2006.)

To further this revolutionary idea, Haughey appointed three senior civil servants to lead the project – Seamus Paircéir (former chairman of the Revenue Commissioners), Tomas Ó Cofaigh (former secretary of the Department of Finance) and Maurice Hogan (retired deputy secretary from the Department of Finance) – and in conjunction with the Industrial Development Authority they began the process of enticing international investment to Ireland.

The attractiveness of the IFSC to overseas operators was dependent on a number of different aspects. Firstly, it was declared a tax haven, and thus could successfully compete with other offshore tax havens including the Isle of Man and the Channel Islands. Secondly, locating in Ireland would give access to European Union markets that these other tax havens could not. Thirdly, competitiveness with Luxembourg, also an EU member and therefore attractive to investors, was assured through the provision of an exceptionally low corporation tax rate of 10 per cent. Understandably the bullishness of this project and its associated incentives incurred the wrath of many other European governments who argued that it was anti-competitive and sharp practice, yet plans proceeded rapidly. It seems that the keen interest the Government took in sponsoring, delivering and regulating the project was critical to ensuring international confidence in its potential (*Sunday Business Post*, 18 June 2006).

The incentives available to potential investors comprised a 10 per cent corporation tax rate (at a time when income tax was larger by a multiple of five) combined with zero commercial property taxes for ten years. Additionally, double tax allowance on rent costs, zero capital gains tax, double tax agreements with a number of countries, and no exchange controls on dealings in foreign currencies were designed to attract three types of financial service activity – fund management, international asset management and dealing in international currencies and securities (*Financial Times*, 11 August 1987 and *Irish Times*, 13 October 1987). To trade at this location and benefit from the attractive deal, financial institutions had to be certified by the IDA following the production of a detailed business proposal. This covered the primary activity of the company, the types of activity for which approval was sought, a development plan covering a minimum of three years, the amount of office space to be occupied, the projected minimum employment level attainable and the extent to which the project would contribute to the Irish economy (IDA Ireland, 1997).

Not everyone welcomed this development with open arms; some people saw it as nothing more than a contrived tax haven that would do little to improve Ireland's prestige in international terms. Barry Desmond of the Labour Party was vehemently opposed to the development, arguing against the speed, lack of thought and lack of transparency that went into it, but also because of the way in which investors were being sought. He argued for 'responsible financial services in this country, not the kind of touting around that has been done to get people to set up in this centre. I suspect there isn't a postmistress in this country, if she made a decent application, who could not get in' (*Financial Times*, 27 September 1988). Doubtless this was in response

66 Demolition of stacks for the IFSC development, 1988. (J. Brady.)

67 Block 1 nears completion, 1989. (J. Brady.)

to the activities of the senior civil servants or 'Three Wise Men' who 'travelled the world with the IDA for the first few years. Getting meetings with the very top people in the international financial services industry and among regulators was no problem to them' (*Sunday Business Post*, 18 June 2006). Although large companies like AIB and Bank of Ireland expressed an intention to apply for licences and showed much interest in the original development, by January 1989 the development still awaited a major international banking house. This may have been because international investors were nervous about the robustness of the world economy following the stock market crash of 1987. Not only was this a problem in itself but it also highlighted the wider danger of depending on international financial service activities for broader urban renewal. This scepticism was confirmed in 1991 when Olympia and York, the developers of Canary Wharf in London, the project on which the Dublin scheme was modelled, declared bankruptcy.

Although an International Financial Services Centre was the aspiration and focus of much government energy, without the backing of the Irish financial services industry this could not have been achieved. Michael Buckley, who had also become a member of the Board of the Custom House Docks Development Authority, described the first four years as 'very difficult … there were few international "takers" and the flag was largely carried by commitments from AIB, [Dermot] Desmond and Bank of Ireland to buy the first three buildings and put their businesses there' (*Sunday Business Post*, 18 June 2006). AIB occupied the first block of the IFSC, totalling 31,586 square metres, in 1990. This was quickly followed by the sale of the south block, IFSC House facing the Liffey, to financier Dermot Desmond (Figure 67) and the occupation of the North Block by Bank of Ireland.

Turning the corner: the IFSC post-1990

By late 1989 it appeared that global economic confidence was beginning to re-emerge and a turning point was reached in Dublin. Firm commitments had been received from 39 institutions to commit to operations at the IFSC. Of these, 26 were non-Irish and included major international players from the US, Europe and Japan such as Chase Manhattan, Citicorp, General Electric, Nixdorf and Sumitomo Bank. In a show of support for the overall project Citibank, who had operated a small business in Dublin since 1965, took a strategic decision with the advent of the International Financial Services Centre to relocate to this new development.

68 The Liffey vista by 1990. (J. Brady.)

69 Nearby development – Irish Life, Abbey Street, 1990. (J. Brady.)

In 1993, the bank and its staff of 80 moved to IFSC House at Custom House Quay and became the largest tenants, occupying 1,300 square metres on the third floor of the building and leasing another 836 square metres in late 1994. In describing why Citibank were so willing to invest right at the start of the project, one representative noted in interview that 'there was a view that this was an area that was going to develop into the top location for financial services firms in the years to come and Citigroup wanted to become a part of the IFSC from the beginning. The plans for the area were impressive, the location was superb in terms of employee access, and the Government was committed to making the centre a success.' A central location at the heart of the city made much more sense for the company than locating in a more peripheral location on the edge of the city, such as Sandyford. This view, took much longer for other international companies to develop, and some considered the cost base of a docklands location, even with the range of incentives in place, still too high. One official argued that a key reason for the slow take-up of space in the IFSC was the high cost of rent which within the development were on a par with London, New York and Tokyo. The difference was that in these global cities business was located in a normal city-centre environment whereas in Dublin they were located in the middle of a building site (*Irish Times*, 30 November 1994). These difficulties were exemplified through the initial difficulties experienced by Dermot Desmond in meeting the costs of acquiring IFSC House. Bought in 1990 for an estimated £30.5 million, average rental charges began at £355 per square metre of space. Following negotiations, rents were revized downwards and, in 1999, Ulster Bank was reportedly paying £296 while Citibank were paying £317 per square metre for commercial space. In contrast to other city centre locations, these rent levels were extraordinarily high, but the quality and specifications of the new buildings were unrivalled in Ireland. With an upturn in the property market in the early 1990s, First National Building Society bought two floors of IFSC House for £16 million, and Irish Nationwide bought another two. Nonetheless, their lack of confidence in the sustainability of the entire development was displayed by the negotiation of a clause that would enable them to sell the space back to Desmond for the buying price if they decided to relocate elsewhere once the tax breaks ended. This kind of bargaining shows the high personal cost of the centre to Dermot Desmond and his determination to ensure its success, recognized by leading politicians at the time and later. The late Charles Haughey acknowledged that 'success has many fathers but the real father of the financial services centre was Dermot Desmond whose concept and idea it was' (*Irish Independent*, 23 May 2001).

70 The George's Quay close to the IFSC had lain undeveloped for most of the 1980s. (J. Brady.)

After an initial halting and uncertain start, construction proceeded more rapidly and in October 1991, an extensive public tendering process resulted in the sale of two more blocks, Harbourmaster Place One and Two. By the end of that year, over 140 companies employing a total of 2,800 people had been approved by the IDA (MacLaran, 1993). Because the completion of office accommodation at the Custom House Docks complex was not keeping pace with the number of licences being issued, many companies continued trading in other parts of the city where they had previously been located, yet benefited from the IFSC incentives. Even though this pent-up demand for accommodation existed and was easily quantifiable, the development of the IFSC, like the rest of Dublin City, was adversely affected by the collapse of the commercial property market in the early 1990s. The biggest blow to the overall vision for the area came when the development consortium ceased building, arguing that the economy had stagnated to such an extent that development was no longer feasible and that speculative construction in such an uncertain economic climate would prove too great a risk. For many months the development remained in jeopardy, as like in other districts, most notably Mountjoy Square at the time, it appeared that half-completed buildings or vacant sites would

71 Dublin Exchange Building, Custom House Docks. (N. Moore.)

dominate the landscape for an indefinite period (Figure 70). This situation was not allowed to continue, perhaps because of the high personal and political costs for a large number of people should the project have failed. Protracted and acrimonious legal wrangling between the Hardwicke/British Land development consortium and the Custom House Docks Development Authority followed as the Authority sought confirmation that the developers were legally bound to continue construction. Development recommenced in August 1993 and shortly afterwards the Dublin Exchange Facility (Figure 71) was opened.

This small building, although not as ostentatious as the larger IFSC buildings on the Beresford Place side of the Custom House Docks, is highly symbolic of the global nature of the financial industry today. It demonstrates the manner in which new technology and communications are driving increased global connectivity, and the new-found ability of international companies to benefit from a range of markets irrespective of time or location differences. FINEX Europe is the European branch of the New York Cotton Exchange (NYCE) and is located within the Dublin Exchange (FINEX) building. At this location, cotton, frozen concentrated orange juice, potatoes and other commodities, as well as interest rate, currency, and index futures and options are traded. The hours that this facility operates in Dublin combined with a day and night trading session in New York ensures that the FINEX markets trade around the clock. A FINEX broker can provide price quotes in all major centres, day or night. The scheme attracted by Forfás to Stack L, one of the smaller original warehouses on the site, is a 'custom-built, floor-traded, open-cry currency futures exchange' (CHDDA, 1994), conjuring up mental images of scenes from films like Wall Street or from the business news channels and contributing significantly to the emergence of a new global image for the docklands.

The opening of this unique kind of enterprise provided a new display of confidence by both national and international companies in the financial services centre and led to an expansion in construction activity that continued through the Celtic Tiger economic boom during the second half of the 1990s. It is difficult to know whether the economic turnaround led to the acceleration of interest in the IFSC scheme or whether, as is perhaps more plausible, the development of the Custom House Docks project and the attraction of such large quantities of international investment capital actually drove the Celtic Tiger prosperity. Whichever is closer to the truth, new buildings constructed between 1994 and 1997 at Harbourmaster Place and George's Dock were leased extremely quickly. One of the most important decisions taken during this period as part of a new Planning Scheme was the enlargement of the Custom House Docks Area. This extended the regeneration remit of the Custom House Docks Development Authority to the wider area. To further this objective, the former An Post sorting office building and areas of Connolly Station surplus to the requirements of CIÉ, were designated for financial or other related office services adjacent to existing IFSC buildings. These and the former Sheriff Street postal sorting office were the subject of a new tendering process for the rights to develop.

Following the earlier difficulties between the original Hardwicke/British Land consortium and the development authority, it is not surprising that although they submitted a proposal to expand their development rights to the new site, their tender was unsuccessful. The winning plan, produced by an alternate developer, Brian Rhatigan, comprised six additional office blocks of 1,858 to 3,716 square metres – Custom House Plaza – and a total of 200 car parking spaces to complement the IFSC. Almost immediately, in May 1996, the first building, Custom House Plaza 1, was leased to ABN Amro, a major Dutch banking corporation (Figure 72). They had been one of the first institutions to commit to the IFSC in 1989 and this decision to extend their operations signalled a high level of confidence in their investment decision.

72 ABN Amro Building, Custom House Docks. (N. Moore.)

The decision to support Rhatigan's commercially-focused development plan began to indicate a radical policy and tactical shift by the CHDDA. Almost overnight, the previously unacknowledged emphasis on pure commercial development, as opposed to the mixed-use scheme originally envisaged, was cast in stone and the purpose and meaning of the redevelopment scheme changed. The warnings of politicians that had emerged the previous decade in debates over the future of the Custom House Docks gained sudden, new urgency as commercial development became the primary concern. While continued support of the IFSC was clearly economically sensible, it resulted in the sidelining of wider regeneration objectives. What may have sounded like hollow warnings in the economic and political context of the 1980s, '… do not have it as a huge block of aluminium and glass … do not have many parts of it used for only six or eight hours a day' (O.J. Flanagan, in Dáil Debates, 9 March 1982) suddenly appeared an accurate prediction of the future.

Financial Services: the engine of the Celtic Tiger?

Since the expansion of the Custom House Docks Area in 1994 and the completion of the second phase on the original site by Brian Rhatigan, a large number of additional office buildings have been constructed in the next phase of the development outside the original Custom House Docks boundary, IFSC II (area 4 in Figure 65). Since 1997 the new Dublin Docklands Development Authority, which is discussed later in this chapter, have further supported the growth of the centre through marketing it as the backbone on which the rest of the development will be delivered. While a significant number of large office buildings now dominate the waterfront on the proposed National Sports Centre site, the second phase of the IFSC has succeeded in broadening land use in the area quite successfully. Along with Commerzbank, Citibank were again one of the most speculative banks in terms of their involvement in the second phase or expansion of the IFSC. As well as a full service branch, Citigroup in the IFSC hosts a regional processing centre known as the Dublin Service Centre. The service centre provides operations and support for the customers of the corporate and investment bank. The Dublin office is one of the largest in Europe and hosts a mix of many different and diverse Citigroup businesses. In July 2005, Citigroup Ireland invested €10 million in establishing the first R&D Centre of Excellence by any bank in Ireland and the first of its type for Citigroup. The processes and services provided include, funds transfer, funds administration, investigations, customer and client services and outsourced treasury. There is also an insurance underwriting operation in Dublin and technology support for other European offices. Citigroup currently employs 1,400 people in Dublin and the name change that took place in July 2006 from Citigroup Ireland Financial Services plc to Citigroup Europe plc is another indicator of the success of this investment in the city. This expansion from a base of 80 employees in 1993 is indicative of the surge in growth and the importance of the financial services sector to the wider urban and national economy.

Following the publication of the new planning scheme in 1994, criticisms have particularly focused on the sidelining of social elements of the Custom House Docks project, yet the economic benefits of the development were far beyond what had previously been imagined. By the end of 1996 a total of £250 million had been invested in the area, hundreds of companies both domestic and international had established operations in Dublin under IFSC licensing arrangements, and 3,000 jobs had been created. Exchequer gains included £643 million in tax revenue between 1988 and 1995. As most companies benefited from a 10 per cent corporation tax rate as part of the incentive package that encouraged them to establish operations in Dublin, this indicates that £6.43 billion profit was generated in the first seven years that the IFSC was in operation. Economically, the potential of this location continues to grow with an annual tax return at the end of 1999 of £300 million having grown to an estimated €700m by the end of 2002. According to the Central Bank of Ireland monthly statistics, international bank assets located in Dublin by April 2007 amounted to €454 billion. Initially perceived by some as non viable and conceived as a pipe-dream in the 1980s, Haughey and Desmond's vision of Dublin as an internationally recognized financial centre in the new global economy has been realized. The original IFSC and IFSC II combined have produced 184,000 square metres of new commercial space in the city, covering 15.8 hectares.

In terms of the overall economic benefits of the development, the original development consortium received very favourable returns on their investment, with over £50 million profit generated to 1997, split between the development consortium and the CHDDA. Hardwicke Ltd, a wholly-owned Irish company, made over £10 million on the project, of which a significant proportion was paid to the Exchequer as tax revenue (*Sunday Business Post*, 20 October 1996). Today IFSC-licensed companies originate from scores of countries, including the major core economies of Germany, the United Kingdom, United States and Japan. This is a clear indication of the international profile of the centre and the successful attempts of the Irish government to benefit from the hyper-mobility of investment capital in an increasingly global economy. For the first time in 2005, the funds industry in Ireland broke through the one trillion euro (one thousand billion) mark, of which some €633bn represented funds domiciled in Ireland. More than 430 international operations are approved to trade in the IFSC, while a further 700 managed entities carry on business under the IFSC programme. In 2005, 1,485 net new jobs were created in Ireland's international financial services (IFS) sector. According to *Finance Dublin Yearbook 2006*, total employment in the three core sectors of banking, funds and insurance at the end of December 2005 was 19,095, a growth of 8.4 per cent from the same date a year earlier.

The impact of this project in general economic terms was felt as early as 1996 when the volume of business within the designated area was such that lending by IFSC companies had outstripped the Irish domestic banking total (*Irish Times*, 4 October 1996). One of the original goals of the IFSC was to create jobs for Irish university graduates and to pre-empt the so-called 'brain-drain', the emigration of young and highly educated Irish graduates to North America and Europe in particular. By the mid-1990s it appeared to be successful in its objectives. The importance of retaining this well-educated

73 Aerial view of docklands, 1994. (DDDA.)

workforce within the country is highlighted by Nicky Sheridan, managing director of Oracle Ireland. While, 'business-friendly policies have created the fastest growing economy in Europe ... this, coupled with a smart, young, flexible, multi-lingual workforce, means that Ireland offers successful multinationals like Oracle a uniquely dynamic environment from which to trade globally'. Employment opportunities in the regenerated area were also diversified by new developments at the nearby East Point Business Park in the software and electronic technology spheres, a designated Enterprise Zone developed with generous fiscal incentives in a similar manner to that at the Isle of Dogs in London docklands. Global companies such as Cisco Systems, AOL, Oracle Corporation, Eircom, Vivendi Universal, United Air and Sun Microsystems have located major operations at this site, attracted by rents that are approximately 40 per cent less than comparable city centre office accommodation and service charges that are about 33 per cent of the cost of other locations. Thus commercial regeneration in docklands has two distinct profiles, high quality office space accommodating major financial operations at the core of the IFSC with less expensive and more extensive space available within the East Point Enterprise Zone.

Benefits and costs of the incentive-led approach

While generally perceived as a total economic gain, some commentators and critics have highlighted the need to also examine the economic costs of the IFSC development. The key criticism has centred on the manner in which incentive schemes encouraged companies already trading in Ireland to relocate to the CHDA (Malone, 1996) or enterprise zone. A prime example is the relocation of Sun Microsystems from Percy Place to East Point in 1996. No measure has been developed to ascertain the number of 'real' new jobs created as opposed to relocated opportunities or to assess the detrimental impact of the incentive schemes on other areas of the city that did not have similar supports in place. The significant commercial development that has recently been undertaken in other parts of the city centre, particularly around Harcourt Street, as well as in the wider city-region including City West, Park West and Stillorgan Business Park suggests that, despite favourable docklands incentives, growth in this waterfront area has not had negative repercussions on the city and instead has played a key role in the development of economic clustering and critical mass.

In terms of possible economic drawbacks, income has been foregone to the Exchequer because of incentive schemes. A KPMG report produced for the Government in 1996 estimated that within the original Custom House Docks Area alone, the cost of the incentives was between £100 and £140 million (€127–€177 million). This does not take into account the cost of schemes within the East Wall and Grand Canal Enterprise Zones. Nonetheless, the argument could be made that this is miniscule in comparison with the tax-take from corporation and personal income tax generated by employment at the IFSC. At a broader spatial scale one negative impact has been the international disputes that have arisen over the use of tax incentives. Early on the German, Danish and Swedish Governments disputed with the Irish Government the tax position of companies from those countries trading within the IFSC. They feared that locating within this semi-tax haven was being used as a means of tax avoidance (MacLaran, 1993). Certain operations have come under close scrutiny as fears arise that many companies used the IFSC as a brass plate address to launder foreign currency. New EU rules and stricter regulation suggests that this is unlikely to be still an issue.

What is indisputable is that the International Financial Services Centre, which began life in the 1980s with the dual mandate of regenerating a decayed part of the urban core and stimulating an ailing national economy, has

74 Aerial view of docklands, 1997. (DDDA.)

undoubtedly re-shaped the financial and policy landscape, as well as the built environment in the docklands, in its twenty-year history (Figure 75). But the original Master Plan of 1987 envisioned a much broader based redevelopment programme with the IFSC as only one, albeit core, element of it. So, what happened to the other parts of the development that aimed to 'exploit fully the site's fine waterfront setting with style, flair and imagination, pursuing excellence of design and quality of construction' offering a 'unique opportunity to add to the social, economic and environmental qualities of Dublin' (Custom House Docks Planning Scheme, 1987, p. 3).

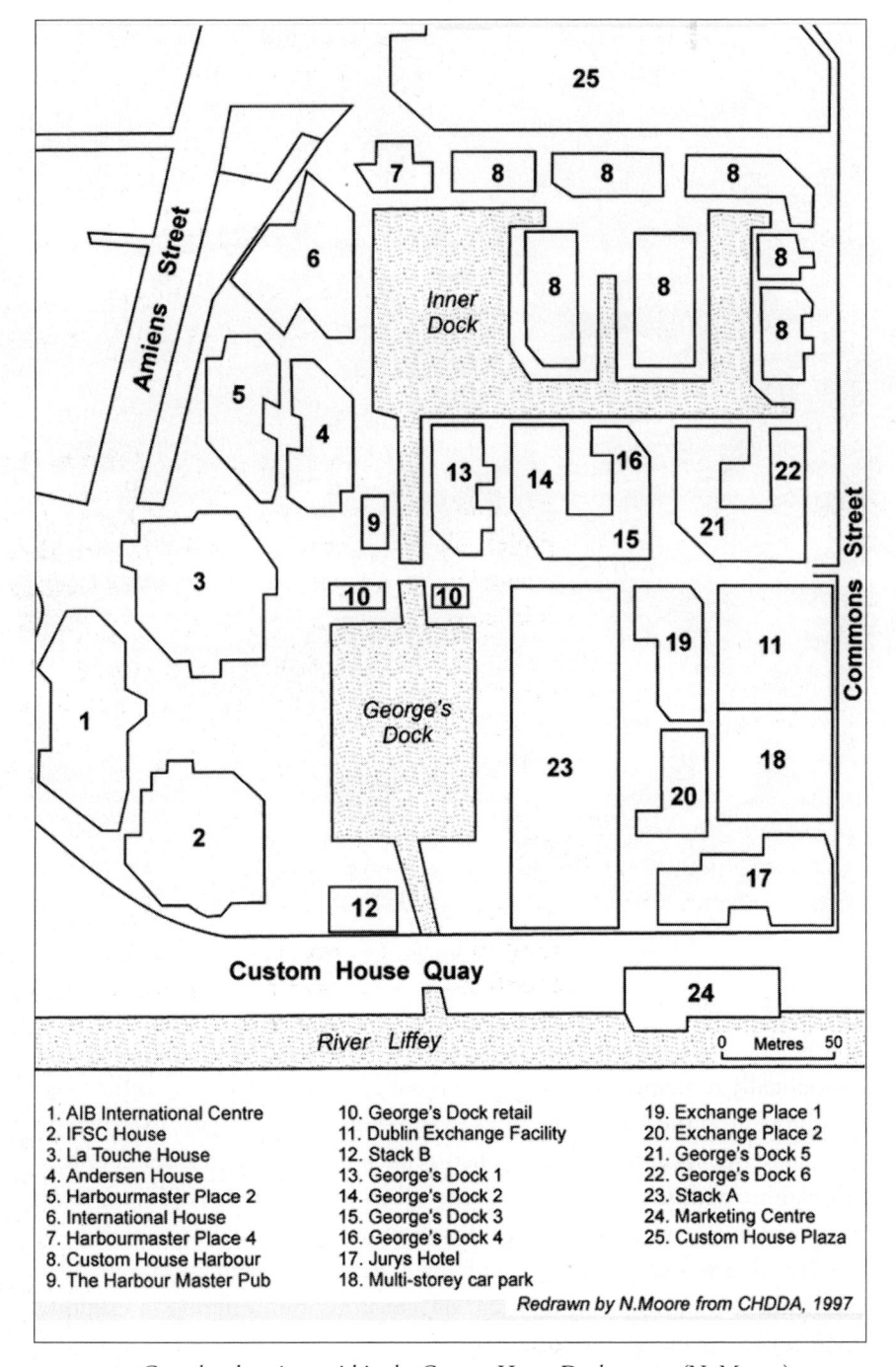

1. AIB International Centre
2. IFSC House
3. La Touche House
4. Andersen House
5. Harbourmaster Place 2
6. International House
7. Harbourmaster Place 4
8. Custom House Harbour
9. The Harbour Master Pub
10. George's Dock retail
11. Dublin Exchange Facility
12. Stack B
13. George's Dock 1
14. George's Dock 2
15. George's Dock 3
16. George's Dock 4
17. Jurys Hotel
18. Multi-storey car park
19. Exchange Place 1
20. Exchange Place 2
21. George's Dock 5
22. George's Dock 6
23. Stack A
24. Marketing Centre
25. Custom House Plaza

Redrawn by N.Moore from CHDDA, 1997

75 Completed projects within the Custom House Docks, 1997. (N. Moore.)

Other elements of the Custom House Docks scheme

Even though many of the world's top financial institutions established commercial activity in the Custom House Docks during the early 1990s, very little ancillary development was completed to complement the extensive office-based function. The mixed-use development of town houses, retail outlets, a food centre, three museums, a waterfront hotel, and public recreation areas envisaged by Benjamin Thompson in 1997 with an anticipated phased completion date of 1992–6, had not even begun as late as 1995. Most of the delay can probably be accounted for by the earlier dispute between the builders and the CHDDA over the viability of non-commercial development. Yet others might argue that it was only in response to pressure being brought to bear by commercial interests and in particular major anchor tenants within the IFSC, that four retail units and a bar/restaurant were opened at Mayor Street Bridge in early 1996, the most successful venture in the complex being the Harbourmaster Pub (Figure 76). Rather than contributing to general regeneration and bringing vitality to the area, these developed primarily to facilitate financial services employees, illustrated clearly through the opening hours of the facilities. The small Spar retail unit and sandwich bars opened only from Monday to Friday and the Harbourmaster Bar initially opened for business lunches only. Occupying the nineteenth-century dock offices, the building has been renovated to retain its maritime ambience. The bar now opens on weekends and has subsequently expanded to extend its restaurant facilities.

In contrast to the large shopping areas zoned in the original planning scheme, the only retail facilities on the original site are a small convenience store that is adjoined by a launderette facility and two small restaurant/coffee shops. Shortly after these facilities were opened, Jury's Hotel Group plc acquired and opened Jury's Custom House Inn along the quayside in September 1996. This 234-bedroom hotel was the third largest in Ireland when it opened and was developed at a cost of £11.25 (€14.3) million, including the construction of a multi-storey car park (Figure 77). It was the first facility in the area servicing a broader consumer base and was arguably the first part of the development scheme that succeeded in attracting non-IFSC employees to the area.

What most of the facilities constructed at the IFSC had in common was a targeted market base, aiming their services directly at the new, 'cash-rich time-poor' residents and IFSC employees. As relatively few of the commercial tenants have changed since the original opening of the retail units a decade

76 The Harbourmaster Bar, Custom House Docks. (N. Moore.)

77 Jury's Custom House Inn, Custom House Quay. (N. Moore.)

ago, it would appear that they have proved a valuable investment. Even though the pace of such developments in the former Custom House Docks Area was much slower than initially anticipated, they far exceeded regeneration in the social or cultural realm. By 1996, a clear gap or contrast had emerged between the intentions of the original planning scheme and the development priorities of those with direct control. This could be explained in large part by the embracing of a neo-liberal, market-driven approach to development in Dublin that clearly emulated similar developments in many cities across Europe and North America at the same time. The kind of strategies adopted in the drive towards attracting international investment in an increasingly competitive global environment resulted in profound disappointment among many groups that had hoped for major social improvements at the local level.

The changing residential structure

Like the rapid internationalization of the financial sector, from the late 1990s onwards, a new style of inner-city living began to emerge in Dublin as a result of regeneration. This closely resembled mainland European preferences for living accommodation rather than the standard suburban property that had proved the norm since the foundation of the State, discussed at length by McManus (2002). Those companies that recognized at an early stage the new demand for inner-urban residences generated by employment growth made huge inroads into control of the residential property sector, among them Zoe Developments. Following their success in other designated areas and in response to the initiation of interest in the Custom House Docks Area, this company constructed the first private residential development of modern times along the waterfront in 1989. Although a small private development had been constructed in the Liberties earlier in the 1980s, Zoe Developments were the first company to engage in large-scale private sector development within the inner-city that had until then been almost entirely dependent on public sector housing provision. They have maintained an interest in the docklands to the present day with its subsidiaries Daninger and Fabrizia developing the Gasworks site on Barrow Street. The large Brooks Thomas site between Spencer Dock and the Point Depot was bought in 2005 by this group and it is likely that there will be significant development here in the coming years.

However, in the early 1990s, Zoe Developments were pioneers in the south docklands. The new complex they developed close to the East Link Bridge outside the designated area emphasized the international flavour of inner-

78 Fisherman's Wharf, Ringsend. (N. Moore.)

urban waterfront living. It played on this perception through its name, Fisherman's Wharf, conjuring up images of districts in other waterfront cities like San Francisco and Sydney. Located opposite the Point Depot, on the south side of the East Link Bridge along York Road this was the first speculative building in the area as the developers foresaw a Dublin version of the docklands boom that was beginning to affect other cities. Fisherman's Wharf comprises thirty-six units in total, thirty three-storey houses with a garage and six two-storey houses (Figure 78). In the context of contemporary apartment complexes, it was not a very large development but it proved effective in testing the water for the developers. The promotional material issued to entice potential investors cited proximity to the Custom House Docks redevelopment, the Point Depot refurbishment, the potential development of the National Sports Centre and water amenities associated with the river as being among the many advantages of living in the area. Ironically, after a dearth of private sector residential activity in the area for a number of decades, most units were sold within a few hours of the launch, making the development, at the time, the fastest selling complex ever in Ireland and giving the first signal that a change in urban residential choice and patterns was beginning to emerge.

The outstanding success of this development which facilitated what appeared to be a pent-up demand for high quality private residential units along the waterfront close to the city centre, resulted in the immediate construction of a similar complex, Alexandra Quay, adjacent to the original development. This was later extended in 1991 following the acquisition of adjoining lands by Zoe Developments. Similar to Fisherman's Wharf, some units in the Alexandra Quay scheme overlook the River Liffey while others overlook an enclosed courtyard entered through clearly defined security gateways. The creation of such defensible space through inward-looking urban design is one of the key criticisms levelled at inner-urban residential schemes constructed since the late 1980s in Dublin and many other cities internationally. This style was something that eventually became replicated all over the city from Smithfield to the Liberties and even within the original Custom House Docks site, where there had initially been such high hopes for new public housing, an expectation that had perhaps lingered since the 'Gregory Deal'.

Far from the plans envisaged by the local community at the outset of the Custom House Docks project, the largest individual residential scheme constructed within the docklands area in the early 1990s was the private enclave of Custom House Harbour. Originally part of the 1987 Planning Scheme, it was not until the publication of a revised scheme in 1994 that construction began. This was partly due to the fact that residential development was not perceived to be as lucrative to the development consortium as office and other commercial construction, a total contrast to contemporary patterns of construction and property values. As mentioned earlier, the *á la carte* implementation of the development plan became a significant bone of contention between the development authority and the subject of a legal dispute. To settle the disagreement, the new contract signed between the authority and developer in 1993 included a clause that all the necessary infrastructure for residential development would be put in place as soon as feasible.

A site around the perimeter of the Inner Dock was identified as the most suitable for the location of 333 new apartments, in seven six-storey blocks, with a range of apartment sizes that would maximize the water views (Figure 79). One-bedroom units ranged in price from £49,950 to £57,950, while two-bedroom units occupied 50 to 75 square metres at a selling price of £60,700 to £76,950. Six two-bedroom penthouses within the scheme sold for £88,950 each and by the end of 1995, a total of 267 units on the former coal-shed and timber yard had been completed. All remaining residential units within the Custom House Docks Area were completed by April 1997 and the

79 Custom House Docks apartment development. (J. Brady.)

80 Security at Custom House Docks. (J. Brady.)

residential element of the previously derelict site was now valued at over £22 million. In hindsight, the anger and disbelief with which the Dublin Port and Docks Board greeted nationalization of the area in 1982 appears entirely justified, given that the compensation they received at the time was worth less than half the residential development value alone in 1997. Their disappointment was undoubtedly compounded by the fact that their plans, while criticized at the time for being too commercially-oriented, were in retrospect visionary in nature given their similarity to the eventual development promoted by a variety of government agencies and constructed just ten years later.

'Socio-economic cleansing' at Sheriff Street

Whatever the extent of tension between the Dublin Port and Docks Board and central government over the Custom House Docks site in the early 1980s, this was just the first in a range of strained institutional relationships that emerged in the docklands into the 1990s. Yet the difficult relationships between various statutory agencies paled into insignificance in the light of the significant divisions that emerged between the traditional residents of docklands and central government over the future of the Sheriff Street flat complex described above. The Sheriff Street flats, although not part of the original designated area, and a substantial proportion of the area retained for the proposed National Sports Centre were included in the Planning Scheme published in 1994, heralding a new phase in the ongoing dispute over this part of the city.

Part of the argument for demolition lay in the fact that between 1981 and 1986, the number of people resident in the complex fell from 1,259 to 831. A decrease of just over one third, this must be seen in the context of general inner city depopulation. The changing objectives of urban policy in the second half of the 1980s were explicitly aimed at reversing this trend right across the city. The number of units occupied was relatively high even though actual population had fallen, and in May 1989 when the Minister for the Environment ordered de-tenanting of the flats, 90 per cent of the 400 dwellings were still occupied. While the long-term community had recognized the problematic nature of the area and had campaigned for demolition and re-housing in the area, this had been undertaken in a very different political context and pre-dated the Urban Renewal Act. As soon as the new development on the Custom House Docks site began to take shape, the community foresaw a significant opportunity for *in situ* upgrading and requested that

81 Juxtaposition of Custom House Docks development and Sheriff Street flats, 1996. (DDDA.)

refurbishment of the existing flats rather than demolition should become a priority (Figure 81). The strategy behind this change is difficult to unravel but may have been driven by a number of different factors.

The first was that new development is likely to have been perceived as an opportunity to upgrade the area in general. Mick Rafferty, a key community activist argued quite reasonably that 'the Sheriff Street flats have to become part of the development. At the moment the people of Sheriff Street stand in the way of the authority's plans – it's hard to attract people to invest beside a slum but this provides a golden opportunity to upgrade the area and the standard of living there by employing them on the site and building decent housing for them' (*Irish Times*, 14 July 1987).

A second reason why local residents shifted their opinions on the best way to move forward may have been an understandable fear that if the flats were demolished the entire community would be broken up and people would be re-housed in different parts of the city. This issue was raised by Ted Johns, a

82 Sheriff Street, 1997. (N. Moore.)

long-term activist in London docklands, who visited Sheriff Street in July 1988:

> On a tour around the Sheriff Street area, I found that the community there were desperate to have the area redeveloped – just as we were ten years ago, when the London Docklands scheme was taking off. But, like us in the Isle of Dogs, I fear that their hopes and dreams will soon turn to bitterness and disappointment as their community is broken up.
>
> (*Irish Times*, 7 July 1988.)

Interestingly just as the Government and CHDDA were visiting and gathering ideas from similar waterfront redevelopment projects initiated elsewhere, similar networks of experience or practice were also being built up more informally by local residents groups. A report produced jointly in 1989 by the CHDDA, Department of the Environment and Dublin Corporation recommended that the flats be de-tenanted and some sold off for private-sector refurbishment. In a letter of 9 May 1989 to the City Manager, Frank Feely, a Department of the Environment official stated that:

> I am directed by the Minister for the Environment to inform you that following consideration of the report of the working group on the Sheriff Street Flats complex by the Government, they have agreed that proposals should now go ahead for de-tenanting the flats and their demolition, and for the disposal of the cleared site. The Corporation should now go ahead with the preparation of an outline plan to implement the decision over a two to three year period ... As previously indicated the Sheriff Street Flats should not be included in the recently notified tenant purchase scheme for local authority flats.

This was a significant development. The local authority traditionally responsible for public housing were now being directly instructed by central government, through Pádraig Flynn the Minister for the Environment, how to manage the housing stock. The opportunity for residents to become more involved in upgrading their local area or even receive the chance to purchase their flats if privatization was considered the best way forward, was denied to them. The political ramifications were so serious that Bertie Ahern, as Minister for Labour, requested written clarification from Pádraig Flynn over the future of the tenants. The latter reassured Ahern on 19 May 1989 that the majority of the tenants 'have agreed to the demolition of the flats on the understanding that they will be provided with better housing accommodation'. A palpable change in mood by the local community from their initial optimism to a more confrontational approach was epitomized by Gerry Fay, local activist, who warned that:

> Maybe they thought when they started dealing with us that the community would be dislodged and moved on, but the reality is that we've fought hard to stay. People who are moved out of the Sheriff Street Flats are going to be re-housed nearby. The dock authority will ignore the local community at their peril.
> (*Irish Times*, 29 December 1989.)

While demolition and re-housing was a significant concern for the local community, controversy further developed due to the differential manner in which residents were treated by the authorities. At a public meeting that was designed to appease widespread disagreement but probably created more, the Minister for the Environment stated that long-stay residents (those living in the area longer than 15 years) would be re-housed in the immediate locality,

83 New public housing at Sheriff Street. (J. Brady.)

while shorter-stay residents would be re-located due to financial considerations (Kyne, 1989). A number of tenants chose to be re-housed outside the area and were provided with a choice of residence accordingly, but the majority wished to remain in the district. As so much investment was projected to occur on an adjacent site (around the Custom House Docks) and there appeared ample room within the original planning scheme to build private sector apartments, the ministerial contention that all residents could not be re-housed in the Sheriff Street area proved particularly explosive. In a sub-committee report to Dublin Corporation in May 1990, the chair of the group established to examine this issue, Tony Gregory, demonstrated that there was more than enough land available to house all the tenants in the St Laurence O'Toole parish as 'there were 13 sites in the area, which could accommodate 253 new houses and 30 houses to be renovated providing another 50 units'. Under instruction from central government, Dublin Corporation used its compulsory purchase powers to acquire several nearby sites. A number of senior citizen dwellings were constructed on Oriel Street, and 119 new units were constructed on Sheriff Street Lower on the northern side of Sheriff Street (Figure 83). The British-based St Pancras Housing Association also developed twenty units for Dublin Corporation. This caused substantial controversy at the time as the Irish Council for Social Housing (ICSH) had not been consulted nor had the St Pancras Housing Association become a member of this umbrella group. Local TD Tony Gregory rejected these complaints citing the fact that the ICSH had failed to deliver rapidly enough on some other sites already assigned to them in the city centre (*Irish Times*, 12 February 1994).

Yet all of this debate became irrelevant as following the extension of the Custom House Docks Area to the Sheriff Street flats site in 1994, no doubt remained as to the intentions of the development agencies. In the introduction to the new Planning Scheme, the future of the Sheriff Street flats was clearly laid out: 'It is the stated intention of both Government and Dublin Corporation to sell off a significant section of the southern side of the site with a view to seeing the existing blocks developed for private apartments' (CHDDA, 1994, p. 4).

With the completion by the City Council in early 1995 of the new units on the northern edge of the site, two blocks of flats were demolished and later that year a further group of new houses were completed adjacent to Spencer Dock. Dublin City Council prepared for disposal of the flats site through the publication of an *Urban Renewal Development Brief* for Sheriff Street in July 1995, entitled the Mayor Street Lower Urban Renewal Project. The aim was to facilitate the development of the former National Sports Centre site, described in the 1994 Planning Scheme as 'a flat, featureless site, which has largely been cleared of a considerable amount of derelict building … [representing] a blank canvas to the authority in determining its future use and visual appearance'. In the brief, Dublin Corporation stated that 'the Corporation is supportive of the overall scheme prepared by the Custom House Docks Development Authority and the Mayor Street Renewal Project will form part of this wider urban rejuvenation scheme'. By agreement with the local communities this was to include an amenity between the new public housing and the soon-to-be constructed private dwellings. This was a critical issue for local groups, including the North Wall Community Association. In a letter to the CHDDA of September 1994, they argued that:

> We have got to address the issues of space for our very large and very young community. During the course of the house building programme large tracts of land and community facilities have regrettably been lost. We are seeking a substantial park and play area as part of the development plan for the remainder of the flats site. We regard this as absolutely vital. It is our sincere hope that the development plan will have as its

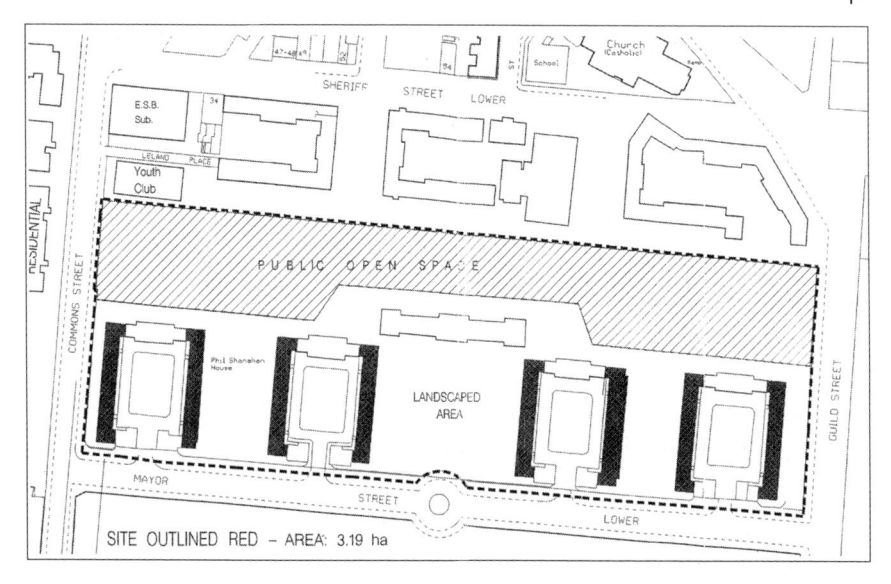

84 Outline diagram of redevelopment showing public park.
(Dublin Corporation.)

main concern the need to address the above issues, the need to act as a means of amalgamating the old and new Communities together to form one strong integrated Community. It would be a tragedy and a disaster if the plan was destined to divide, even in a subtle way, both Communities.

The northern part of the development site was designated as a public open space that would remain under the control of the City Council. A separate, enclosed landscaped area would be provided in the interior courtyards of the private flats, formed by joining two blocks of the refurbished older flats together through the addition of new blocks along the northern edge to create a U-shape (Figure 84). On the development plan produced by the CHDDA, the public park appears as a dividing or separating feature rather than as an integrative, shared space as the Dublin Corporation development brief suggests. The reality of what later emerged is much closer to the CHDDA perception with a very small playground separated by high fencing from the new private apartment blocks (Figure 85) providing a buffer between the two residential groups.

By late 1996 de-tenanting had been completed and the 3-acre site was eventually sold to a private developer, Chesterbridge Ltd, for £4.3 million

85 Playground as a buffer zone, Sheriff Street area, aerial view, 2005. (DDDA.)

86 Playground as a buffer zone, Sheriff Street area, 2005. (J. Brady.)

(€5.46 million). One measure of the difficulty that the Corporation had experienced in trying to sell the site, given its previous relatively undesirable reputation, was the agreement that payment for the site would only be made by developers to the local authority when the apartments went on sale. It would also appear, given the significant gap between the proposals in the development brief and the reality of the area post-development, that the developers used their upper hand with the local authority to change the original specifications and remove some of the proposed elements.

Unlike the official proposals that had proposed private sector refurbishment of eight blocks of flats, by February 1998 the last block of flats was being razed. On this site now stands, Custom House Square comprising 620 apartments and a wide range of retail units including a beauty salon, coffee shops, Spar, an office supply store, florist and various eateries (Figure 87). Not only has the land use been completely changed, the entire morphology of the area has been re-written. A new award-winning public open space, Mayor Square, now links the development on both sides of Mayor Street (Figure 88). New offices have been constructed along the riverside of the street with private and social housing, a new college and some extra retail activities. The new residential structure and college are discussed in the next chapter in relation to the changing socio-demographic and economic profile of the wider area, but on the former flats site, the public space has been squeezed as the new S-shaped apartment blocks meander along the public-private boundary. The divisive nature of the small strip of land to the rear of them has lived up to the fears of the North Wall Community Association, and is a tangible reminder of the tension that has emerged between the two different social groupings.

Some of those who have been re-housed in the vicinity argue that the demolition of the flats resulted not only in a changing physical landscape, but the fragmentation of the 'community spirit' that previously existed, a detrimental rather than benign impact of the redevelopment project. The depth of emotion generated by the debate was evident from the residual strength of feeling still felt about the process. In a broader discussion around social regeneration in docklands, Gerry Fay of the North Wall Community Association emphasized that 'none of us could afford to buy the apartments that were built on the site of the old Sheriff Street flats. That's a form of economic cleansing. People who grew up in the area cannot afford to live there. It has the same result as ethnic cleansing: you get out' (*Irish Times*, 25 February 2000).

The division between the long-term residents and new residents, particularly in the apartments developed in the IFSC around the Inner Dock, has been

87 Custom House Square, Mayor Street. (DDDA.)

88 Mayor Square and environs. (N. Moore.)

reinforced by the retention of a large wall along the northern edge of Commons Street. This separates the Sheriff Street Youth Club, constructed in March 1995 by the CHDDA as a form of community gain, from the original private apartments at Custom House Harbour. Rather than removing it, as was the original intention in the 1987 Master Plan for the area, the wall had in fact been heightened with netting (Figures 89 and 90). The divisions and different perception within a community that is often considered entirely homogeneous is displayed by the fact that some of the longer-term residents did not see the retention of the wall as a major problem. Some argue that 'it's to keep us apart from the yuppies. They've made their own ghetto in there', while others understand that 'they need it. If the wall wasn't there those nice flats would be wrecked once the gurriers got at them' (*Irish Times*, 30 September 1997).

Thus while perhaps the regeneration undertaken by the CHDDA did not significantly improve the quality of life for long term residents of docklands, it did alter the relationship between them and their immediate surroundings. The physical environment and services developed within the original IFSC site became increasingly alien to a community that required basic intervention across a range of social issues including education, employment and community-support infrastructure. For many, their new 'neighbours' became representative of, and were benefiting from, the products of global capital and conspicuous consumption that was beyond the reach of the majority. But this is nothing particularly peculiar to the Dublin setting, as this growing trend towards increased social polarization seems to be a characteristic of areas that quickly move from industrial to post-industrial activity. In many recently regenerated dockland areas, significant social polarization has emerged, manifest in wider income gaps between groups and variable lifestyle and employment opportunities.

The limited nature of the remit adopted by the Custom House Docks Development Authority from 1994 precluded much being done to either recognize or stem these difficulties. What did become increasingly apparent, through the 1990s as the economic success of the IFSC grew was that the needs of deprived inner city communities are not best, or even adequately, served by the private sector alone who fail to deliver important community facilities. Some might even argue that the needs of the city as a whole were poorly served socially and culturally by the eventual outcome of the Custom House Docks development.

While the scheme was an indisputable success in an economic and physical context, many of the projects were a struggle to develop or were in fact simply

89 The 'wall' between Sheriff Street and the IFSC, 1999. (N. Moore.)

90 The 'wall' between Sheriff Street and the IFSC, 2005. (J. Brady.)

removed from the plans. One notable omission was Stack A, the historic warehouse that was envisaged as the social and cultural hub of the project. Only now, twenty years later is this element at the forefront of the development agenda for a range of reasons discussed in some detail in a later chapter. Even the original sponsor and key promoter of the regeneration, Dermot Desmond, made his disappointment with the limited nature of the finished Custom House Docks project very clear in an oral hearing over the Spencer Dock development in February 2000. While acknowledging the major economic success of the IFSC, he commented:

> Less satisfactory has been the failure to provide the diverse range of facilities set out in the original plans … none of these are there. I have gone on the record over the years with the Docks Authority and the relevant Ministers expressing disappointment at these shortcomings. I have battled against departures from the original development ideal. We have failed in one of the key objectives which was to make the IFSC a vibrant, social and cultural area … What developers say they will do and what they actually ultimately deliver is entirely different.
> (*Irish Independent*, 23 February 2000.)

This perception is not unique and is indeed reflective of a much wider set of concerns regarding the outcome of the Custom House Docks project. The lack of progress made by the CHDDA on the social infrastructure promised in the initial plan was disappointing, in contrast to the high-end restaurant and consumer service facilities that proliferated. Social change and polarization became an increasingly critical issue and the activities of the CHDDA began to be read as a form of exclusion by omission rather than by any overt action. By the late 1990s, the lack of amenities had become a major concern for locals and was a situation that even business leaders remarked upon. Bill Cunningham, Managing Partner at Coopers & Lybrand (now PriceWaterhouseCoopers), stated that:

> I suspect that the impact of the development on the local community was probably initially negative. But hopefully this is now levelling itself out. There is certainly no real hostility although the development doesn't seem to have benefited the local community to date.
> (*Sunday Business Post*, 20 October 1996.)

Local communities felt increasingly let down and understandably powerless against a State agency – the Custom House Docks Development Authority – that was increasingly pushing a very clear neo-liberal agenda in support of private development. While this may have been inescapable in the climate of recession that characterized the late 1980s and early 1990s when all efforts focused on attracting private investment to risky locations, this approach became increasingly untenable as the property market began to boom. Changes in the broader political context also meant that this approach to development was becoming increasingly inappropriate. Just as neo-liberalism and Thatcher-style policies of privatization had influenced the development of the initial plans for regeneration, the increasingly apparent drive towards a more partnership-based model of politics had a later influence on the regeneration project. Epitomized by New Labour in Britain, the policies of the Clinton Presidency in the United States and successive national partnership programmes in Ireland, the changing political climate in the mid-1990s, led to palpable changes in the approach to urban revitalization. Social inclusion became a key government priority, developed through a range of new structures. These included the establishment of Social Inclusion Units within local authorities and a new approach to large-scale regeneration such as the Integrated Area Plan projects in major urban centres, as well as a new plan for the docklands.

A need for change: the establishment of the DDDA

In trying to understand the patterns of development within Dublin docklands over a decade from the mid 1980s, it might be reasonable to suggest that the emphasis of renewal shifted so rapidly from the mixed-use to a much greater commercial orientation because of the changing economic climate. During the 1990s, Ireland experienced an unprecedented economic boom with an average growth rate of 9 per cent in GNP between 1995 and 2001, in contrast to most other European countries that experienced annual growth levels of between 4 per cent and 5 per cent. The result in Ireland was a shift in the economic structure of the country to emphasize the importance of foreign direct investment and the attraction of service activities. As already suggested, it could be argued that the economic success of the IFSC drove the engine of the Celtic Tiger growth. Others suggest that this was just one outcome of a much broader shift in government policy that favoured growth. Leading economist, Brendan Walsh of University College Dublin, suggests that the economy had receded to such an extent in the late 1980s that it had nowhere to go but up and that it was both internal and external factors that generated the boom. Externally, the availability of EU structural funds for major capital projects and the general increase in US investment in Europe was a positive thing and at a national level, the Government created the necessary conditions to foster growth through the extension of a favourable tax regime to this emerging financial services sector. He also suggests that while the Irish Government showed foresight in supporting this sector, the IFSC in the mid-1990s was effectively subsidizing an already booming business sector.

As this occurred, the demands from the social partners to build a fairer, more inclusive Irish society that would address the challenges of poverty reduction and social inclusion grew more difficult to ignore. It is within this wider political-economic context that the operation and achievements of the CHDDA came under closer scrutiny. A report commissioned by the Department of the Environment and produced by KPMG in 1996 to benchmark the impacts of urban renewal policy highlighted many limitations in its effectiveness. The kinds of mixed-use development and heterogeneous social profile that was intended for areas subject to renewal programmes had not emerged. The Government also recognized that the kind of spin-off benefits promised to trickle down to marginalized communities had also failed to happen and nowhere was this more apparent than in the growing divide between the new Custom House Docks area and the surrounding communities. The 1996 Budget statement delivered by the then Minister of Finance Ruairí Quinn, a public representative whose constituency included part of the south docklands, announced the intention of the Government to redress this issue by adopting a strategic approach to the greater docklands area. This 1300-acre (526 ha) expanse comprises one-tenth the size of the city between the two canals. Physically much of the area comprised wasteland and brownfield or old industrial sites, while in social terms, above-average concentrations of unemployment and other social problems persisted. Quinn, whose Labour party colleague, Brendan Howlin, was Minister of the Environment, had worked informally with him to generate the initial ideas and assess the feasibility of this new approach for a substantial part of the capital city. Once they had an agreement in principle, Brendan O'Donoghue who was Secretary General in the Department of the Environment drafted the bulk of the legislation. The fact that Quinn as junior minister at the Department of the Environment in 1983/84 had already worked with O'Donoghue no doubt helped the emergence of a close working relationship between the small grouping. In a written response to a series of questions put to him, Ruairí

Quinn indicated his belief that O'Donoghue was probably one of the key people in ensuring the introduction of the extended docklands project stating that 'he was enthusiastic from the beginning. Indeed, I believe he played quite a role in drafting the formal letter of response from Brendan Howlin to myself.'

Although the announcement of a strategic approach was broadly welcomed, the manner in which the key issues could be addressed produced much greater concern. A special task force was established to report to the Government on the most appropriate strategy and they concluded that a new development authority with a specific multi-purpose remit going beyond the economic objectives of the Custom House Docks Development Authority was necessary. In his speech, introducing the Bill to Seanad Éireann, the Minister of the Environment stated that:

> The influence of the urban renewal experience during the last ten years is evident in the mandate to be granted to the new authority. While the authority will have an important function in improving the area's physical environment, its remit will extend far wider than that, encompassing the sustainable social and economic regeneration of the area.
>
> (Seanad Debates, 27 February 1997.)

In May 1997, the Dublin Docklands Development Authority (DDDA), subsuming all functions of the CHDDA, was established by an Act of the Oireachtas with an organizational structure designed to guarantee more democratic control and participation than under the CHDDA. Gerry Kelly, Director of Social Regeneration with the DDDA has contrasted the activity of the new authority with that of its predecessor stating that 'the [new] docklands body has a very strong social input' (Oram, 1998, p. 33). A docklands council of twenty-five members, many of whom are local community leaders, was established with a role in recommending particular proposals to the seven person executive board. This represents a forum through which community and other interest groups develop and present long-term visions for the docklands and is very different from the attempts of the CHDDA to establish a tokenist Community Liaison Committee. The statutory inclusion of local communities in the decision-making process has meant that socio-economic objectives are now key priorities. The most recent plan highlights this very clearly by stating that 'the inclusive approach set out in the Act provides for a more democratic process, inviting communities to participate in and contribute to the development of their areas. This provides a broader spectrum

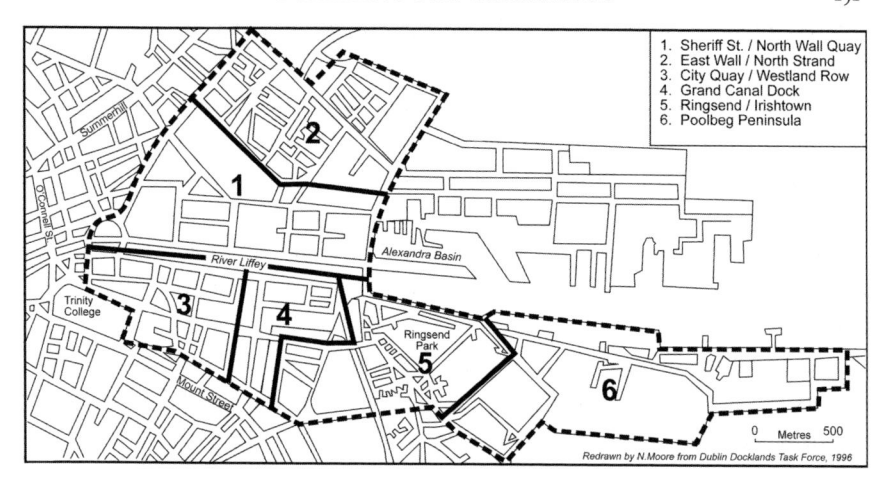

91 Scale of the extended docklands area, 1996. (N. Moore.)

incorporating physical, economic and social considerations, including education, training and employment opportunities' (DDDA, 2003, p. 7). While this is certainly the case, it is also worth noting that there is some disagreement among local community representatives regarding the effectiveness of the Council, a situation that is examined in the following chapter.

Other than the development of new decision-making and organizational structures, one of the other greatest changes with the establishment of the new authority has been a major change in approach to planning the area. Mirroring the approach of agencies in other cities such as Cardiff, which had begun to engage in a process of waterfront regeneration at the same time as Dublin, the DDDA recognized the difficulties of managing an area of this size and diversity. Instead of trying to deal with every issue in one sweep and with a single plan, a strategic decision was made to adopt a sectoral approach. In other words, an overall Master Plan would be produced for the docklands area within which six integrated area plans would be drawn up, one each for the Custom House Docks area; East Wall; Poolbeg Peninsula; Ringsend/Irishtown; City Quay/Westland Row and the Grand Canal Dock (Figure 91).

Because some areas like the Grand Canal Dock required massive infrastructural investment in comparison to other more viable areas such as East Wall and Ringsend/Irishtown, only the most derelict districts became eligible for tax incentives. Had the entire area been eligible for tax incentives, the attractiveness of this location for investors would have resulted in massive negative consequences for the rest of the city and defeated the purpose of regenerating one part of the city only to create a major problem in others.

The 1997 Master Plan for docklands

The overall aspiration of the range of plans developed for the docklands since 1997 has been the creation of a living, breathing and self-sustaining community that fits somewhat with Benjamin Thompson's original aspiration to create 'The City of Man' at the Custom House Docks. Unlike any previous planning document, the 1997 Master Plan identified a set of specific objectives or targets against which the performance of the authority should be judged after their term of office. This move may have been an attempt to reassure local representatives sceptical of the authority, by guaranteeing a measure of accountability right from the start. Indeed a relatively impressive record has been established in a number of key sectors to date. The main objectives of the plan in terms of land use were to ensure the vitality of the district by increasing the residential base of the area by 25,000 over fifteen years, and ensuring a balanced social mix. One of the most innovative elements, that along with a local employment charter proved critical in obtaining legitimacy among those who had been side-lined by the CHDDA, was the introduction of a provision for 20 per cent social and affordable housing in all new developments. The provision was eventually included in response to the concerns of local communities during the drafting of the scheme even though it caused some disagreement. Community activists argued that if property speculators continued to operate in the area, young people would not be able to remain living in their local area and would be displaced by new upwardly mobile residents destroying the existing community base and leading to an unbalanced social mix. Had the community representatives not had the right to veto the plan under the new institutional arrangements, this key concession might not have been achieved. In fact, it might be suggested that this proposal played a key role in influencing national urban policy, given that Part V of the Planning and Development Act of 2000 proposed the rolling out of a similar provision across the country. This provision is obligatory in many European countries, such as Germany for example, but the experience in Ireland had been quite different given our history of private home ownership. Other key achievements during the negotiation of the Master Plan include policies relating to retraining of the local population and increasing levels of educational attainment. The most radical proposal, and again the one that caused much difficulty, was the introduction of a minimum requirement that 20 per cent of new full-time jobs and apprenticeships would be reserved for local residents. The record to date of the docklands authority in achieving

92 New DART station at Barrow Street. (N. Moore.)

these and many other social objectives is examined in later chapters, dealing with the changing socio-economic character of this district.

So that these objectives are realized, the key requirement is the attraction of people to the area. In recent years, Dublin City Council have made many attempts to reduce the number of private cars in the city centre, but the major obstacle to achieving this is the lack of an efficient and affordable public transport system. Many areas of the city are very poorly serviced by bus and rail, resulting in a high reliance on private transport. Dublin docklands is one of these locations and the authority has proposed a number of possible solutions to this problem. With the completion of the Dublin port tunnel and the introduction of a ban on heavy traffic in the city centre, much of the heavy traffic along North Wall Quay bound for the port has been eradicated. But traffic calming is just one part of the solution; getting people into the area is the critical issue. The opening of a new DART station at Barrow Street was one of the key objectives of the authority to open up the Grand Canal Dock area (Figure 92) and the recently opened Spencer Dock train station will perform the same function on the north side of the river. Yet, facilitating mobility within the area is a much more difficult issue and there are proposals in the Master Plan for the establishment of a Dockland Area Bus System

(DABS). These and many other challenges have been addressed in a review of the original Master Plan that resulted in the publication of an amended and updated version in late 2003.

At the outset in 1997, it was estimated that the total cost of the docklands regeneration would be in the region of €2 billion, with €444 million pledged by public bodies and the remainder being met through private investment over the fifteen-year lifespan of the plan; the Dublin Docklands Development Authority are now speculating that total investment will eventually exceed €10 billion (DDDA, 2007). Some districts have been designated as Section 25 areas, which means that planning schemes similar to those in the Custom House Docks Area will be implemented. These areas include the existing Custom House Docks Area and North Wall Quay as far as the Point Depot; the Poolbeg Peninsula; and the Grand Canal Basin. What this means is that specified districts are exempt from the regular planning process and that the DDDA can give permission for any development once the proposals are within the terms and fit the overall vision of the Master Plan. In the remainder of the area, along the Liffey campshires (the area between the road and the river, similar to a boardwalk) and within the Royal Canal linear and urban park, the Dodder river linear park, the village centre of East Wall, Ringsend/Irishtown and the City Quay/Westland Row precinct, detailed action plans are being developed.

Managing the future

In co-operation with Dublin Corporation and private sector developers, the Dublin Docklands Development Authority is moving towards achieving a radical re-orientation of the axis of the city even further eastwards and the re-establishment of broken linkages between the city and the waterfront. Sensitive and coherent land use planning is an imperative and for this reason individual plans have been published for smaller-scale areas within the overall docklands scheme. Two of these, the extended Custom House Docks or North Wall Quay area and Grand Canal Dock have Section 25 status, which means that detailed schemes have been prepared for these districts and the docklands authority will act as the planning body. The opportunities and challenges of these major flagship projects are examined in later chapters, with a discussion of current plans for the Poolbeg Peninsula, a unique part of docklands because of its mix of infrastructure and land uses. Currently it is not a Section 25 area, yet it has been subject to some recent interesting proposals that also warrant extended discussion. In the 2003 Master Plan, the DDDA suggested that this area could potentially be covered by the kind of planning regime implemented in the Custom House Docks and Grand Canal Dock, but acknowledged that Dublin City Council have already developed an Area Framework Plan. In fact, the nature of this plan was considered so politically sensitive that publication of it was deliberately postponed, at the request of city councillors, until after the local elections in June 2004, again highlighting the very direct relationship between the political, planning and urban environments.

In the three remaining 'villages' within docklands, integrated area plans have been developed to guide future development. Unlike the Spencer Dock or Grand Canal Dock areas, developers operating here must seek planning permission by applying to Dublin City Council in the usual manner. These plans are non-statutory and represent a vision of how the Dublin Docklands Development Authority would like to see the areas redeveloped, while providing guidance to planning officers regarding planning permission for specific projects. The key objective is to develop a clear vision for development and improvement, while ensuring that the unique existing character of each of these districts is not undermined. The East Wall district in the northeast docklands is already home to a large and well-established residential community, albeit one that is becoming increasingly subject to gentrification. Rather than encouraging new construction, the overall objective in this area is the improvement of the existing environment in order to guarantee vitality and vibrancy. A key way of doing this is through land-use zoning. Industrial zonings have been replaced with residential and enterprise zonings in a number of locations. This feeds into the overall strategy of providing additional residential accommodation in an already well-established zone and there are some suggestions that single, older residents could be encouraged to move out of large, family-size homes if sheltered accommodation is provided in the vicinity.

This may be difficult and add to existing difficulties with residential development. The East Wall community has had strong associations with Dublin port over a long period, but this is beginning to change, as the area has become an increasingly desirable residential location for workers in the East Point Business Park and other nearby locations (Figure 93). Retaining a balanced social profile may be difficult as 'market forces and higher land values are displacing lower value marginal uses in the area, which do nonetheless provide important local employment' (DDDA, 2003, p. 95).

93 East Point Business Park, aerial view, 2004. (J. Brady.)

94 The Wiggins Teape factory, East Wall, prior to demolition, 2005. (J. Brady.)

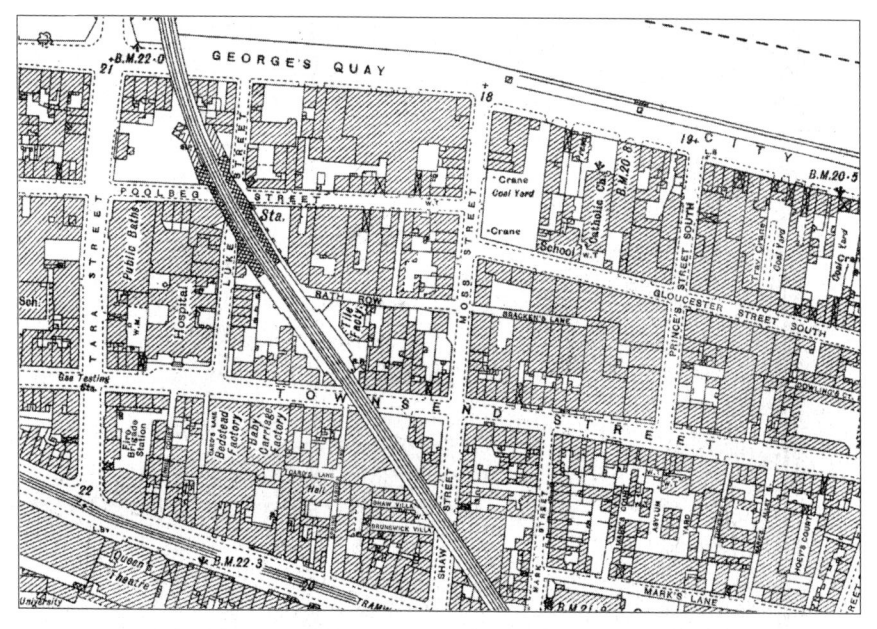

95 The pre-development landscape around George's Quay.
(Ordnance Survey 1:2500 plan, Sheet 18(IX), Dublin, 1939 revision.)

96 George's Quay in 1990. (J. Brady.)

97 The first phase of development nears completion on George's Quay, 1992. (J. Brady.)

98 The completed development on George's Quay – the 'canary dwarfs'. (J. Brady.)

Although the predominant land use, particularly in the vicinity of Church Road is residential, the Dublin Docklands Development Authority is encouraging the development of district-level commercial functions. Many buildings in the area are under-utilized, and others are vacant, detracting from the overall appearance of the district. To remedy the situation, the DDDA in co-operation with the planning authority, Dublin City Council, and local community groups intend to improve the street and pavement surfaces, provide a grant to local residents to improve property boundaries and front gardens, and eradicate 'rat-running' through the area. All of these measures should ensure that the environment becomes more people-oriented and facilitates, rather than deters, social interaction. The objective here in contrast to much of the rest of the north docklands is to promote and facilitate ongoing change rather than re-write the character of the area.

The City Quay/Westland Row area, in contrast, does not have the same cohesiveness in terms of urban fabric and function as the East Wall area, but does provide many more opportunities for development and thus the action area plan is far more detailed. During the decline of the docklands area as a whole, this district suffered dramatically in terms of both physical and social infrastructure. It is typically urban in terms of its functional mix and can be divided into a number of different zonings including residential, commercial, river-based and mixed land use. Given the upsurge of development in recent years, the urban fabric has been significantly altered with the emergence of new 'super-blocks' or large-scale developments where formerly a number of small buildings on individual plots existed. One of the most obvious examples is the George's Quay development that has obliterated an entire street (Figures 95–98).

On Pearse Street, much of the recent construction has been piecemeal with little attention to any kind of overall coherence, which has damaged the continuity and image of the street. This is particularly evident in the vicinity of Pearse Square (Figure 99) and the Gilbert Library (Figure 100), where relatively recent large-scale developments, such as the Holiday Inn, co-exist with much older and often attractive urban fabric. All of the buildings on Pearse Square are protected structures, and contrary to popular perception there are a large number of other protected buildings and streets in this district, including City Quay and Sir John Rogerson's Quay, listed on the Sites and Monuments Record. Physical conservation is thus an important objective of the docklands Master Plan in this district. But it is not just the existing physical form that is often forgotten about. In this area, there is still a significant residential population and as such, the Dublin Docklands Development Authority asserts a wish to work with the local authority to co-fund the development of new focal spaces at a small-scale, improve landscaping, and improve the streetscape at Pearse Street. To date, there is little physical evidence of rejuvenation with the exception of the recent restoration of the Gilbert Library and there are still plenty of derelict and vacant buildings in this area that are ripe for either redevelopment or restoration. This may occur spontaneously as a spin-off benefit if the new cruise liner terminal proposed to accommodate smaller ships on City Quay is eventually developed. How this will co-exist with the extensive new office development that has expanded from George's Quay to the east is not quite clear, but there are precedents for similar developments in other cities, such as New York.

Further east on the south side of the river, Ringsend and Irishtown villages centred on Thorncastle Street and Bridge Street south of the Liffey are some of the most cohesive districts within the Dublin Docklands Area. As both village centres have been designated zones of archaeological interest, the main objectives are the improvement of the existing townscape through new paving, lighting and seating, but there are no plans to engage in wholesale redevelopment. Functionally, Ringsend is a typical mixed-use area with schools, shops, churches, small businesses and recreational space catering for the local residents. But, in recent years, older land uses have been subject to increased development pressure as a result of the close proximity of the area to the city centre. The area plan highlights the need to maximize the potential of existing amenities, such as Ringsend and Sean Moore Parks, while developing new civic spaces such as that which has recently been completed at the Library close to St Mark's Church. A priority will be to increase activity in the parks and, through passive policing, deter the kinds of anti-social behaviour that prevent more vulnerable residents using them. It is intended to increase the residential density close to these parks and thus re-instate them as key features in the physical and social landscape of this relatively old urban district, and the development of new apartments on Pigeon House Road is probably the first steps in achieving this goal.

A decade of redevelopment

Although the 1990s was generally accepted as a time when Ireland, as a country, experienced the most dramatic and rapid changes in its long history, nowhere epitomizes this as well as Dublin docklands. In particular, the

99 Pearse Square. (J. Brady.)

100 Gilbert Library and environs, Pearse Street. (J. Brady.)

Custom House Docks area that began as a derelict, problematic site in the mid-1980s had come to represent the new Ireland – embedded in a global economy, increasingly cosmopolitan and demographically heterogeneous by the late 1990s. While it came to embody all of the positive aspects of the Celtic Tiger economy, it has also in many ways demonstrated very clearly the underbelly of that growth, in particular the emergence of a 'dual city'. While spatial or geographical proximity between a range of social groups has been increased dramatically, in that very different social groups now live 'cheek by jowl' in the extended IFSC/former Sheriff Street flats area, the problem of social polarization became a key concern by the mid 1990s. The establishment of the Dublin Docklands Development Authority with a much stronger social remit than the original Custom House Docks Development Authority must be understood in the context of a changing political-economic environment in Ireland and elsewhere, emphasizing the need to promote inclusion and the development of social capital.

The creation of an extended docklands area in 1997, while a management response to regeneration across a large part of the city, should not be regarded as an indicator of homogeneity within the docklands nor perceived as a desire for standardization across most of the eastern city centre. The character of the docklands area as a whole has been established over many decades through its association with the port, yet distinct villages and communities have emerged over time each with a very different set of needs and priorities recognized in the new area-based approach to development. The multi-layered nature of the planning environment comprized of legislation, strategic and area plans, makes explaining and understanding change a difficult task. Yet, although the changes that occurred during the tenure of the Custom House Docks Development Authority were dramatic, they were very much in line with broader global trends towards the development of cities of consumption and capital, increasingly dependent on international investment and service activity.

Creating a living city: changing directions for Dublin Docklands

> The regeneration of the docklands area will not only be important for the area but it will be of immense importance for the city of Dublin and for the nation and it will raise the city's ongoing renaissance. A city, like any living organism, requires a healthy beating heart. As a strategic part of the heart of Dublin, a healthy and vibrant docklands area will be of enormous value to the city which has seen many other parts of its central area regenerated in recent years.
>
> (Minister Brendan Howlin, Seanad Debates, 27 February 1997.)

By the mid-1990s similar to many other European cities, Dublin was beginning to reap the benefits of sustained economic growth and many parts of the inner city were subject to urban regeneration initiatives. The conditions that generated this attractive investment climate included the increased opportunities for global trade and expansion that had been accompanied by pro-market government interventions across Europe and North America. As the decade progressed another major trend began to emerge as states including Ireland, the United Kingdom and the United States showed an increasing turn to the left in political terms. Although, obviously not 'leftist' in traditional terms, the policies of the Clinton Presidency in the USA, the 'New Labour' Agenda in Britain and indeed the formation of the 'Rainbow Coalition' (comprising Fine Gael, Labour and Democratic Left) in Ireland in 1994 promoted, or at least facilitated, what has been generally termed as the 'turn to community'. Greater emphasis on the social agenda permeated all policy sectors.

The National Anti-Poverty Strategy, promoting a Social Inclusion Agenda, was published in Ireland in 1997 with the aim of poverty-proofing all government policy. This was followed by a number of National Action Plans on Poverty and Social Inclusion (NAPS Inclusion), the first published in 2001, another in 2003 and the most recent one in late 2006. Each of these has been produced as a requirement of an EU Strategy to promote social inclusion in Europe and spread the benefits of economic development. While these strategies are designed to promote economic inclusion and well-being, there is also a broader dimension in terms of promoting participation in decision-making and active citizenship. It is within this broader policy context that the redevelopment of Dublin docklands has taken place within the last decade.

In Ireland, the docklands Master Plan produced in 1997 was one of the first tangible examples of the shift in policy emphasis. While many people had undoubtedly benefited from earlier urban renewal initiatives throughout the city, others were most definitely left behind. Physical redevelopment in the form of the number of cranes on a skyline is often the most obvious indicator of change within a city, but perhaps more dramatic is the social, economic and psychological changes wrought by regeneration programmes. The everyday impact of redevelopment for long-term residents is often unacknowledged by 'outsiders'. As well as opportunities for new jobs and better schools, they must adjust to living adjacent to new and more affluent neighbours, get used to the closure of former small-scale corner stores and the opening of often more expensive chain stores, and put up with increased traffic flows and congestion. While it is widely accepted that Dublin's image has dramatically changed from that of a dirty, old town to an image-conscious, brash and confident European capital, the implications of government redevelopment policies on specific groups have often not been fully considered.

The result of this change is that in the last two decades the social geography of the city has been dramatically re-written as young, affluent twenty- and thirty-somethings move into apartments in areas once considered the least salubrious parts of the inner city and in greatest need. Although the social and demographic profile of the inner city has become increasingly mixed, levels of interaction between different social groups sharing the same neighbourhood remain as low as if large distances segregated them. Balancing economic redevelopment with social policy objectives remains one of the biggest challenges for policy-makers as they try to engineer dialogue across social boundaries. Dublin docklands is a microcosm of the general social changes across the city as a whole, with one exception. Many of the policies implemented elsewhere, such as the provision for a State-wide Affordable Housing Programme in the Local Government Planning and Development Act (2000), were first developed and implemented in the docklands. Perhaps as a result of the turn to community, there has also been significantly more emphasis placed on collaborative planning and consensus building in docklands than in other parts of the city. The result of this and a range of specific, targeted projects has been a social transformation in an area that up to the 1980s was one of the most neglected parts of the state and yet from the mid-1980s became a haven for the type of speculative development that was examined in the previous chapter.

A changing population and place

One of the most obvious themes to emerge so far in considering the evolution of docklands has been the marginal character of this area through time. From its origins until the late-twentieth century, the area had always been a kind of zone in transition. This area was particularly vulnerable to global economic changes. Thriving trade resulted in a bustling port and docklands but the reverse was also true as the vibrancy of the area fluctuated in response to external conditions. This is quite different from the relatively more stable history of the rest of the city, where districts have undergone less dramatic change over longer periods. In 1996, a comparative examination of the socio-economic structure of docklands in relation to the rest of Dublin city and county undertaken by the Economic and Social Research Institute (ESRI) highlighted the extent to which conditions in this sector of the city deviated from the norm. The results were unsurprising, because there was a widespread perception that the docklands was an area in significant difficulty. It showed much higher dependency ratios and lower levels of educational attainment than across the rest of the city. But what has often been ignored is that, like any other large urban area or indeed smaller town, docklands is not and was never a homogeneous zone. Some neighbourhoods were indeed as well, if not better, off than many communities in the suburbs and elsewhere and their population and household structures reflected this. The idea therefore that a 'one-size-fits-all' policy or a blanket description could be applied to docklands is inappropriate. This is clear from discussions with local community groups, who argue that there are many distinct communities within docklands with very different historical trajectories and contemporary needs.

The previous chapter showed that in the mid-1980s mass migration from the inner city resulted in the emergence of a donut-shaped city with growing suburbs encasing an increasingly empty city centre (Brady, 1988; Prunty, 1995). As a significant part of the urban core, the docklands exemplified this trend and between 1971 and 1991, the population of the Dublin Docklands Area showed a pattern of accelerating decline, the most dramatic occurring in the 1970s when a 20.7 per cent decrease was experienced in eight years (Economic and Social Research Institute, 1996). Decline continued through the 1980s but between 1991 and 1996 for the first time in many years, a net increase in population of almost 6 per cent became apparent in the Dublin Docklands Area. Since then the population has continued to expand, reaching a twenty-five year high in the last census (Table 6).

This reversal in population trends was driven by the construction of many new apartment complexes in designated areas throughout the inner city, a tangible demonstration of the effectiveness of government initiatives in promoting city-centre living. This new phenomenon in Ireland, a return to inner-urban living, has also fostered major societal changes by promoting a culture of apartment living among young people in particular.

Table 6 Population change in Dublin docklands.

(District) Electoral Division	1981	1986	1991	1996	2002	2006
North Dock A	1593	1370	1222	1188	1287	1208
North Dock B	4258	4021	3503	3655	3628	3700
North Dock C	2659	2672	2324	2411	3568	4126
Mansion House A	3243	2986	3011	3139	4269	4462
South Dock	3123	2968	2589	3307	3764	5122
Pembroke West A	3674	3233	3070	3292	3241	4276
Pembroke East A	4655	4458	4427	4349	4304	4758
Total DDDA	23,205	21,708	20,146	21,341	24,061	27,652
% change over previous census	-4.1%	-6.9%	-7.8%	+5.6%	+11.3%	+13.0%

(Compiled from Census of Population of Ireland, various years.)

This changing physical structure, a product of property or developer-led regeneration, has had a differential impact on particular neighbourhoods, in some cases because of the crude way in which tax incentive areas were identified. This was nowhere more evident than in Francis Street in the Liberties, during the first part of the 1990s when one half of the street was designated to benefit from incentives, while the other was not. The outcome was a significant gap in prosperity between the two sides of the same street highlighting the danger of drawing the boundaries of designated areas so glibly. Within the docklands zone, similar differences were apparent even before regeneration. While the North Lotts and western part of docklands (North Dock C, Mansion House A) had experienced substantial population gains through the 1990s, by the turn of the millennium some areas were suffering from significant population decline, particularly south of the river. It was not just the intensity but also the nature of change that provided

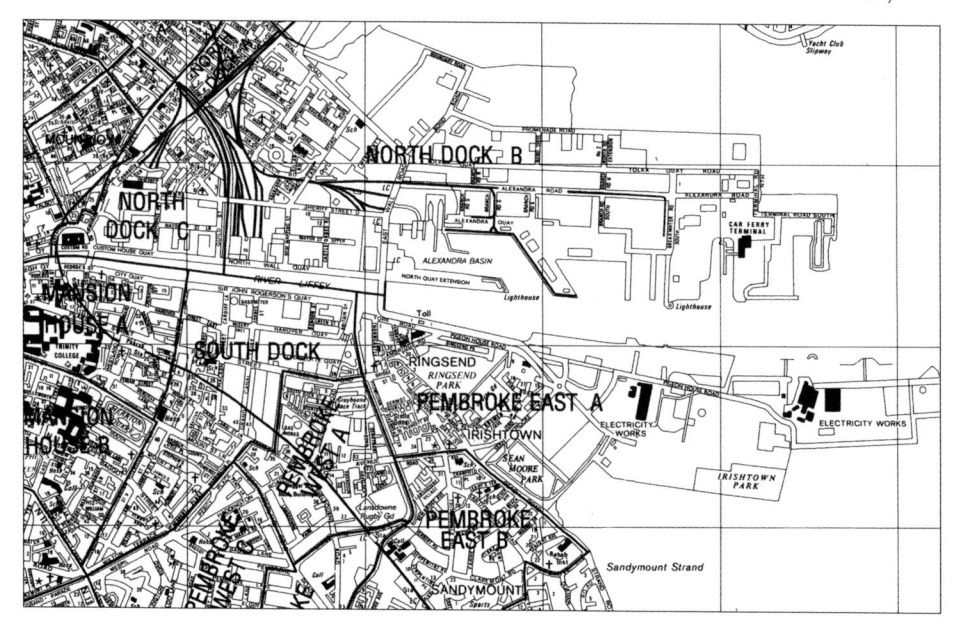

101 Docklands district electoral divisions. (Ordnance Survey.)

important challenges. Most of the new developments in the early to mid-1990s were targeted at very defined population groups. This is very clear in the dramatic transition that occurred in the age profile of docklands residents, particularly the growth in the number of residents aged between 15 and 45, and the decline in population at both ends of the age spectrum (Figure 102).

In characterizations of inner city change, it is generally accepted that a clear indicator of an area in decline is the growth of an elderly population. The latest patterns that have emerged in docklands would therefore suggest that far from being in decline, many parts of this area are demonstrating strong evidence of an urban renaissance. The most striking changes have occurred in the 15–24 year age group and might be partly explained by the construction of new student residences associated with the National College of Ireland. This may create an artificial impression of general population patterns, but it does demonstrate a new vitality within the area. Between 1991 and 1996, the proportion of total population accounted for by this age category increased from 19.21 to 31.95 per cent. One quarter of all houses in this district have been constructed since 1991 and the Census of Population has indicated a major increase in the 25–44 age group. It seems that a significant number of people in their mid-twenties moved into the new apartments constructed throughout docklands in the early 1990s.

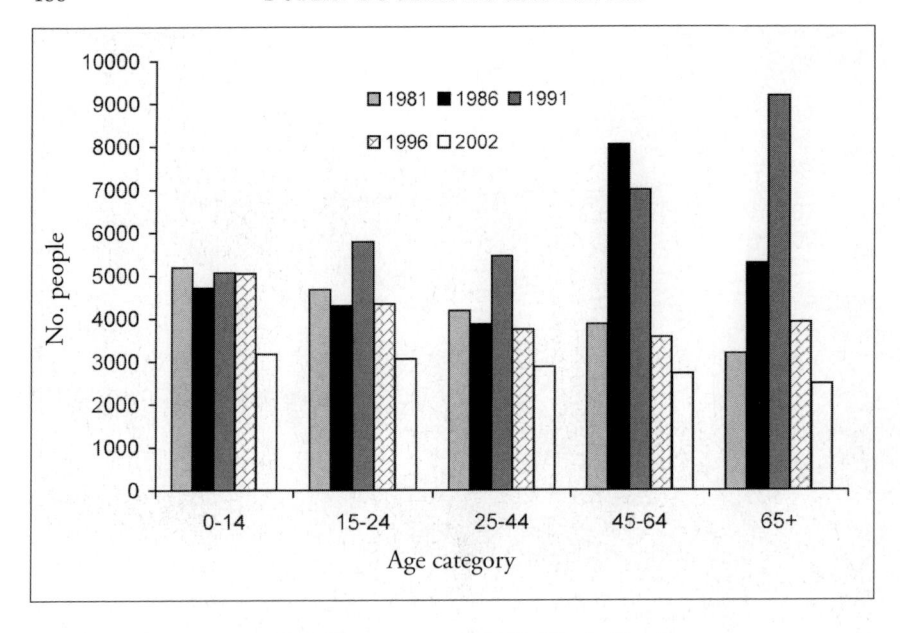

102 Changing age structure in Dublin docklands.
(Compiled from Census of Population of Ireland, various years.)

This youthful population profile has had a major influence on new land use patterns, most apparent in the areas of greatest population expansion, around the International Financial Services Centre and the Pearse Street/City Quay/Westland Row areas. These districts exemplify how changes in the structure of society clearly impact on the urban fabric through the emergence of particular kinds of services and facilities. Aimed at the new, young professional residents the number of launderettes and dry cleaners, convenience food stores, personal services such as the Hair and Beauty Salon on Mayor Street and restaurants and cafés has increased significantly. In Mayor Street on the site of the former Sheriff Street flats discussed in the previous chapter, businesses like Insomnia, the Bagel Factory, Bendini & Shaw, O'Brien's Sandwich Bar, Aya, and the Swedish Food Company have all opened up new ground-floor food outlets. These cater not just to the financial services employees but also to new residents and students, signalling the newfound desire for a café culture in Irish urban society and highlighting the increasing internationalization of consumer demand (Figures 103 and 104). Marks & Spencer have constructed a *Simply Food* store, while Mace and Spar are both significant presences in the area. Gradually the north docklands are becoming a microcosm of the city reflecting but also driving social and cultural change

103 Café culture in Dublin docklands. (N. Moore.)

at a very fast pace. This was caricatured in a recent RTÉ television drama, *The Big Bow-Wow*, set in docklands and providing a very definite image of the emerging city:

> This is twenty-first century Dublin – a boomtown. It's a city of noodle bars, boy bands, Saturday night drug habits, extortion, corruption, fast living, self-consumed singletons. They have money, run businesses, turn the wheels of the town. Some are ambitious, clawing their way up; some are just muddling through. It's an ensemble piece. The central characters are like an urban family, thrown together by geography rather than biology.
>
> (www.rte.ie.)

This radical change in the cultural and social environment contrasts with the more low-key description of leisure time activities that characterized the south docks in the 1920s and 1930s:

> There was a dancehall up in Irishtown, it was called Johnny Pedlars. That was before the flats were built up. There was a big field up there.

104 New commercial outlets on Mayor Street. (N. Moore.)

> It was called Gleesons Field and there was this little hall in it. That was the local entertainment. Mainly the picture houses and that little dance hall.
> (St Andrew's Resource Centre, 1992, p. 45.)

Through the 1970s and into the early 1980s, social life within the city was constrained by the unemployment problem. The degree to which this blighted entire neighbourhoods and districts was nowhere more evident than north and south of the river along the docks. Average unemployment was high and labour force participation rates low in the docklands, yet very clear distinctions in employment characteristics became evident even within the Dublin Docklands Area. Black spots, most noticeably North Dock C which included Sheriff Street, had chronic problems with unemployment peaking at 58 per cent in 1991 when the overall averages were 30 per cent for docklands (Figure 105). Though unemployment was a massive national problem and gave Ireland a reputation for educating the young to use their talents elsewhere, in very few other places within the country or city were the levels quite as high as in docklands. Much of the significant increase in unemployment in the first half of the 1980s in docklands illustrated the immediate impact of port

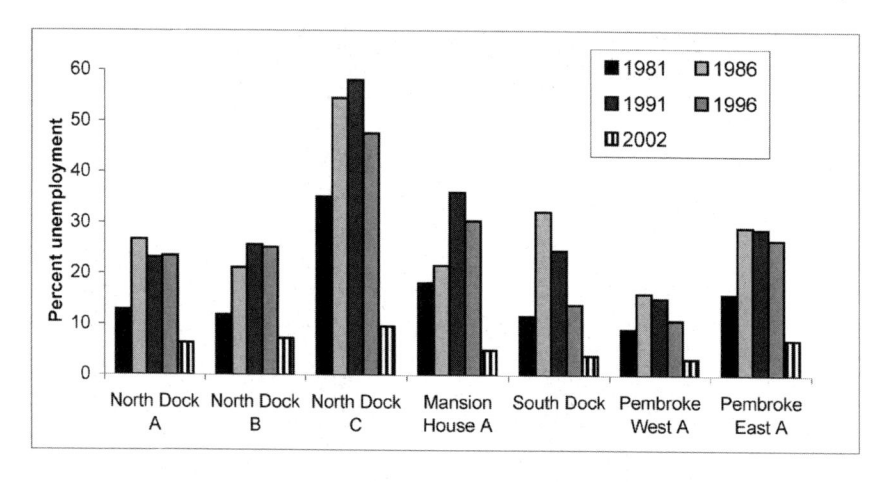

105 Changing unemployment rates in Dublin docklands.
(Compiled from Census of Population of Ireland, various years.)

lay-offs and the skills deficit which precluded alternative employment being found.

More surprising though is how quickly this pattern has been turned around. The Government might point to the activities of the Custom House Docks Development Authority in attracting employment into the area, but this explanation fails to see the nuances in the population statistics. Since 1991 and more so since 1996, the number of middle-class young professionals in the area has grown following an aggressive marketing campaign for docklands. This has had a significant impact on reducing unemployment statistics as many new residents are exceptionally highly educated and skilled. The general situation has improved yet some local community groups would argue that for long-term dockland residents, unemployment is still a significant problem. The radical reduction in overall unemployment figures is more to do with the zero unemployment level among newcomers than a miracle cure for moving the long-term unemployed back into the workforce. The activities of the Dublin Docklands Development Authority since 1997, in conjunction with other agencies and institutions, have specifically focused on this issue and resulted in a marked improvement. Nonetheless, the lack of progress in addressing a range of social issues during the tenure of the Custom House Docks Development Authority produced a landscape of suspicion and mistrust between many agencies and local residents. This mistrust was compounded by the kind of marketing and promotional material being produced by the agency which very much focused on the desires of potential residents rather than the needs and significant difficulties that were being faced by the existing community.

Promoting the 'new docklands'

As early as 1987, the first year of its activities, the CHDDA spent £202,786 on marketing, promotions and publicity in an attempt to kick-start redevelopment and confidence in perhaps the most marginalized district of a then marginal city. The aspirational environment described in a range of promotional material encapsulated and exemplified the potential of the bright new era about to dawn on docklands:

> It is a warm, calm September evening. The highly paid executives in the Financial Services Centre are still at work – their VDUs giving out the latest on Wall Street. At the Liffey's edge the tanned and fit members of the Custom House yacht club are tying up their craft and are strolling leisurely to the dockside pub for a pint or G&T. The kids are not yet back at school. The culture vultures are on their third museum – in the Dublin section they are still not over the shock of what the city was like when it had vacant sites. At the heliport a Ryan Air courtesy helicopter arrives with some more tourists. A limousine whisks them to their luxury hotel. In the apartments a successful young barrister has just arrived home via a vaporetto from the law courts up the quays. She sits on her penthouse balcony admiring the spectacular view of the mountains. As she sips her Campari soda she wonders if the Bunuel movie is playing at the Screen on the dock.
>
> (CHDDA, 1987.)

While it is easy to see in retrospect the poetic licence of this portrait of docklands on a number of levels, not least of which is the time-poor nature of contemporary society as a result of such successful employment growth, all of the marketing drives for docklands since 1987 have been highly targeted programmes. Each of them has promoted and continues to sell the area as a dynamic, futuristic sector of the city in which to work, live and play. Although expenditure on marketing, promotion and publicity by the DDDA amounted to €1,602,565 in 2005, the tone of the publicity has changed and much of it is now designed to showcase community achievements and public events. The costs incurred in marketing seem extraordinarily large, yet the docklands authority is no different from many other area-based or urban development authorities at a global scale. With increasing mobility of international investment capital and greater competition between and within cites, re-imaging, marketing and civic boosterism have become crucial governance functions. The goal of all of these agencies, whether it is in New York, London or Dublin, is the same – to attract upwardly mobile residents with significant disposable incomes who traditionally would not have considered residing in the city centre as well as the multinational corporations with significant resources to invest.

In docklands, the early publicity in the late 1980s was designed to re-image or 'imagineer' the city (engineering an image), in a similar way to places like Pittsburgh in the USA where the local authority attempted to shake off the old industrial image associated with the steel industry through advertising and the development of slogans including 'Pittsburgh: Stronger than Steel'. The idea of many of these campaigns is to market the city on its new physical structure and future prospects rather than on history. It is a deliberate decision by those in a marketing role to wipe the slate clean and present a sanitized maritime ambience or a romanticized version of urban living. The result in many cities is that their docklands areas have been transformed from traditional working class sites of employment to middle-class urban playgrounds, one prime example being Baltimore in the USA.

Yet a major problem is the inability of these marketing campaigns to deal with the needs of pre-existing communities. Through the early 1990s in Ireland and elsewhere, significant tension emerged between the new residents buying into these kinds of marketed lifestyles and long-term residents whose history and place in the community were being ignored or destroyed. As the previous chapter highlighted, local residents initially welcomed redevelopment within docklands as a chance for self-improvement, but subsequent actions and the way in which development progressed succeeded in alienating a large number of long-term residents. By the mid-1990s, both private and public sector activities were viewed with suspicion and generally opposed by locals. While photographing apartment construction in Ringsend, one pensioner resident in the area approached me to say: 'You're not building more of those yuppy apartments are you? We've enough of them already.' Incentive and property-led regeneration became viewed as a threat, rather than opportunity, and was perceived as State-promoted gentrification. New residential complexes in the area, and indeed in many similar areas subject to renewal programmes around the city, have juxtaposed existing underprivileged communities with the new 'yuppies', vividly illustrating and underlining the divisions and inequalities of post-industrial capitalism within modern Irish society. So while social distance is increasing, and recent reports by Conference of Religions in Ireland (CORI) and other commentators testify to the fact that the Celtic Tiger boom has actually reinforced inequality, geographical distance is simultaneously being compressed (Allen, 2000; Bartley & Treadwell Shine, 2002). This has become obvious not only in the Custom House Harbour and Custom House Plaza projects, but also throughout Ringsend, Pearse Street and along the quays. Until very recently community development was not an obvious benefit of rejuvenation. Redevelopment initiatives in Dublin could be held up as textbook examples of gentrification:

> a physical, economic, social and cultural phenomenon. Gentrification commonly involves the invasion by middle-class or higher-income groups of previously working-class neighbourhoods or multi-occupied 'twilight areas' and the replacement or displacement of many of the original inhabitants.
>
> (Hamnett, 1984, p. 284.)

Although the new plans have sought to locate new exclusive developments adjacent to affordable and social housing, the desire to maintain physical distance in line with social distance has resulted in existing communities and the public being physically prevented from accessing new developments in the docklands area. This latter point is one that marketers are increasingly promoting (Figure 106). One of the major selling points, according to estate agents Hamilton Osborne King, of an apartment within Custom House Harbour sold in 2001 was 'the high profile security within the development'. This pre-occupation with protection has resulted in the emergence of gated communities and what has been described as 'defensible space' (Newman, 1972). This ranges from overt security personnel patrolling the area to high-security access points to many of the residential developments. Even Dublin Corporation in the Mayor Street Urban Renewal Brief argued that one of the reasons they extended the physical boundaries of the Sheriff Street flats zone for developers was to make the site 'more attractive by providing additional private and defensible space for the adjoining flat blocks'. This emphasis on security resulted in one of the most contentious issues between the old and new residents, described in the previous chapter, the retention of the 'Berlin Wall' divide on Commons Street.

For Sale by Private Treaty

The Bailey, Custom House Harbour, I.F.S.C., Dublin 1

Section 23 Tax Credit £68,000

This well designed apartment (c. 600 sq ft) is situated on the fourth floor in the secure environment of a perimeter block with a usable suntrap balcony overlooking the water.

A large open plan living room with large windows cleverly adds to the wonderful sense of space. Truly a most desirable pied a terre.

- **Allocated surface car space**
- **High profile security within the development**
- **Passenger lift**
- **Good quality carpets & curtains included in the sale**
- **Electric gold shield heating**
- **High specification kitchen**

Accommodation

Entrance Hall	4.5mx1.1m 14'9x3'6	Hotpress. Water heating by electric dual immersion.
Lounge	5.2mx3.8m 17'0x12'6	With sliding glazed door to balcony 3.7mx1.5m (12'0x5'0)
Kitchen	2.4mx2.2m 18'0x12	Range of fitted presses, worktops with tiled splashbacks, 0stainless steel sink, Indesit four plate hob and oven, extractor hood, washing machine, fridge
Bedroom 1 (left)	2.9mx2.4m 9'6x8'0	With wardrobe and dressing table with mirror
Bedroom 2	4.0mx2.6m 13'3x8'6	With fitted wardrobes and shelved press
Bathroom	2.4mx1.7m 8'0x5'6	Bath with hand shower, pedestal washbasin, wc, strip light

Price Region
£190,000 to include carparking, carpets & curtains.

Viewing by Appointment

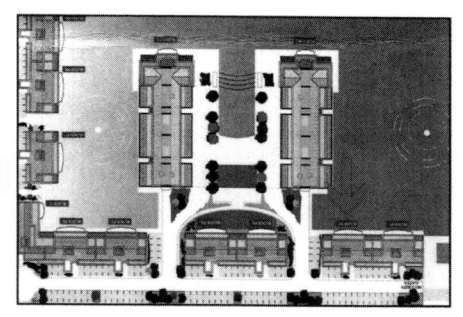

106 Apartment sales leaflet, 1998. (HOK.)

This is an extreme example; more subtle control of space is also exercised through the extensive CCTV coverage of many districts. Similar to many other cities, public space in Dublin is becoming increasingly controlled through electronic monitoring, creating very obvious geographies of exclusion and inclusion. Nonetheless, the most significant problem facing the Dublin

Docklands Development Authority on their appointment was obtaining legitimacy and building a sense of trust with local communities. The sidelining of community interests and needs during the original regeneration at the Custom House Docks site created a very difficult environment and a significant challenge for the new development authority.

Regeneration and 'community capacity'

In contrast to the earlier dependence on the State to intervene in the area, local activism is a key feature of everyday life in docklands today and the development of an effective active citizenship agenda has become a priority. While there have always been a number of local advocates in the area, most notably Tony Gregory and his associates, the fragmented nature of the docklands, its many communities and their differing needs made it difficult to define a clear social agenda. As in other areas undergoing such significant change, inter-community integration and empowerment began to emerge in the late 1990s as a side effect of other actions. One of the most controversial proposed developments in the country in recent years, the original Spencer Dock project discussed in the following chapter, acted as the spur for seven dockland communities to form an alliance in 1999 to safeguard themselves from high-rise development. Labour Councillor Kevin Humphries remarked at the time that:

> They are becoming more pro-active and coming through with proposals telling the corporation and elected representatives what they want for their areas. They aren't just sitting back and waiting for the next development proposal to come along. It may not be a free education, but it's given them the confidence to go out and learn.
>
> (*Irish Times*, 7 September 1999.)

The response of local communities to the new docklands project launched in 1997 was varied, from those who welcomed a chance to draw in investment for social infrastructure such as the East Wall Development Council to a more 'muted approach ... because for many people it was hard to see what the final outcome would be' (Ruairí Quinn, TD). In other quarters the reaction was much more negative, particularly among those communities that had dealings with and had been let down by the Custom House Docks Development Authority. They were mindful of the lack of consultation that had previously taken place in relation to the future of the north docklands during the tenure of the CHDDA and the failure of the project to impact economically to any extent on the local community. In 1993 of 800 construction workers employed on site, only 50 were from the north inner city area (Benson, 1993). Few attempts had been made to retrain the existing population with the skills necessary to gain employment in the new industries of banking, insurance and other financial activities. Approximately 12–15 four-year apprenticeships in carpentry, brick-laying and other trades were offered to local residents between 1988 and 1990, but the community felt let down by the fact that these were offered in other parts of the city, rather than on site at the International Financial Services Centre. No official figures exist to determine the attrition or success rates of the apprenticeship schemes though Kyne (1989) attempted to estimate the short-term employment offered by these schemes between 1987 and 1990 (Table 7).

Table 7 Short-term employment for local residents provided by the CHDDA.

Type of scheme	Locals employed
Construction work (August 1989)	42
Apprentices (Commenced 1988)	15
Apprentices (Commenced 1989)	12
Apprentices (Commenced 1990)	13
Six month construction training	28
Total	110

(Kyne, 1989; CHDDA, 1990.)

Since the establishment of the DDDA a much more comprehensive employment programme has been enshrined in the Local Employment Charter stating that 20 per cent of new jobs in docklands should be offered to local residents first. This provision was a concession that local community representatives fought hard to obtain having initially been offered a 10 per cent quota in the draft 1997 Master Plan. Since then, 120 local jobs have been delivered in the second phase of the IFSC development under the local labour charter. This charter is also currently operational on the Spencer Dock and Grand Canal Dock sites.

Under the Schools Job Placement Programme that began in 1999/2000, 124 young docklanders who have passed their Leaving Certificate exams have been facilitated to enter the world of work and are now employed by IFSC and other dockland businesses. Between 2001 and 2005, 43 apprenticeships in electrical, carpentry/joinery, plumbing/fitting, plastering, painting/decorating and brick-laying have also been provided to local young people wishing to acquire trade skills (Docklands Social Regeneration Unit).

The determination of community groups to ensure that these kinds of benefits accrued to existing residents and that they were involved, if only in an advisory capacity, in determining the future shape of the area has much to do with their negative experience with the Custom House Docks Development Authority. In order to counteract the criticism that the development process, being removed from standard planning channels, was inherently undemocratic, the establishment of a community liaison committee was included in the 1987 Planning Scheme. In the late 1980s, an informal community liaison programme was inaugurated with the North Wall Community Association, Sean McDermott Street Residents' Association, East Wall Residents' Association and the Alliance for Work Forum under the auspices of a community consultation office. Locals regarded this office as a barrier to real consultation, a buffer between the CHDDA and interest groups. Throughout redevelopment, disputes arose between the authority and local communities who argued that consultation only occurred subsequent to a decision or plan of action being decided by the authority. To redress this problem, a Community Liaison Committee was officially established in 1995 and subsequently the Inner City Trust was founded. Set up in partnership by IFSC companies and the CHDDA, the Trust provided grants of £350,000 to local community projects including the new Sheriff Street Youth Club; when compared with the bright new dawn promised in 1987 for this area, this initiative pales into insignificance. One explanation for this is the very low levels of public participation permitted during the lifetime of the CHDDA. One study on public participation in the Custom House Docks redevelopment concluded that the opportunities for public input into the decision making process were negligible, and the level of participation that was permitted could be classified as 'degrees of tokenism' (Creamer, 1998). Given this context and the experience of regeneration in the docklands throughout the 1990s, it is hardly surprising that the establishment of the new Dublin Docklands Development Authority was met with scepticism and in some cases, indifference. While social objectives were very obviously sidelined in favour of economic renewal during the tenure of the CHDDA, this was inevitable because of the limited nature of that agency's brief. Weak legislation that did not compel the authority nor provide the resources to engage specifically in social and cultural rejuvenation resulted in a very uni-dimensional and relatively mono-functional district emerging by 1997. This had gradually become more and more apparent and was compounded by the report of KPMG Consultants presented to the Government in 1996 that made some hard-hitting statements about the social failures of the Custom House Docks project. They concluded that:

> there has been little benefit to date from the redevelopment (of the CHDDA site) to the neighbouring communities in terms of employment, amenity and facilities. Social problems such as drugs, poor education attainment and marginalized communities remain.
> (Department of the Environment, 1996, p. 91.)

Yet there are some indications that the result of marginalization has been a politicization and mobilization of the local community on a scale unseen in many other areas. As all development outside of the small Custom House Docks site remained under the remit of the local authority, who are legally bound to facilitate public participation, local groups did have the opportunity to input directly into proposals that would affect them. The best example is the large number of objections that were lodged against attempts by the Dublin Port and Docks Board to develop the Liffey campshires (the area between the quay walls and the road formerly used as a site for loading/unloading ships) for office purposes in 1997. Local communities felt that, should the development go ahead, they would be robbed of amenity space and access to the waterfront. This has become a serious issue internationally, most notably in Toronto, where access to the waterfront has been privatized. The objections by locals in Dublin docklands illustrated a newfound sophistication and confidence among local communities. It culminated in them winning concessions such as the local labour charter and generous social housing provision in the Dublin docklands Master Plan when it was finalized in 1997. Ruairí Quinn, who as Minister for Finance at the time was one of the key advocates and instigators of the docklands project, reports that 'when the Council [community and other stakeholder representatives] were reviewing the draft development plan which was prepared by the DDDA, they insisted on the social dimension for the housing component of the plan. It was from this that the requirement for a 20% social and affordable housing content of all residential development was first promoted'.

Since 1997, local community representation has been formalized on the Council of the Dublin Docklands Development Authority which is charged with contributing to and directing development proposals as well as ratifying area action plans. The Council also includes representation from other stakeholders such as the Irish Congress of Trade Unions, Dublin Chamber of Commerce, IBEC, Waterways Ireland and An Bord Gáis. While this was generally considered in a positive light, some members of Seanad Éireann during the second stage of the Dublin Docklands Development Authority Bill questioned how well the membership of the council had been thought out. Senator Feargal Quinn argued that:

> there is no representation of tourism interests which should be at the heart of it. There is no representation from Trinity College, yet a prime objective should be to persuade Trinity to use the docklands as its preferred base for campus companies. On the other hand, there is representation from An Bord Gáis which is unnecessary because at the moment Bord Gáis happens to be the owner of a very large derelict site where the gasometers used to be. The board should, and in all probability will, sell that site and get out. As a company Bord Gáis has no role in developing the economic potential of the docklands. I am not sure that boards must have representation but they should have something that reflects the future of the docklands rather than its past.
> (Seanad Debates, 27 February 1997.)

The Council was constituted as initially proposed and its function is primarily consultative. It does not have any veto over particular decisions and their decisions can be over-ruled by the Executive Board. This ability and desire to participate is something that perhaps sets the docklands communities aside from their suburban counterparts who have not yet had to develop a pro-active approach to safeguarding their local environment. Although the docklands council perhaps appears to provide a democratic façade to the decisions of the development authority, it is still the Minister of the Environment who sanctions appointments from a list provided by local groups. The first Council was appointed by then Minister for the Environment, Brendan Howlin, in consultation with his political colleagues. Ruairí Quinn, local TD and cabinet colleague, recalls that 'there was very close cooperation between Brendan Howlin and myself on the composition of the Board of members who were appointed as well as the Ministerial nominees from the community sector to the Council'. While this could be seen as problematic, the current chair of the Community Liaison Council has remarked on being impressed with 'how little party politics plays a role' in how the docklands structures operate. Nonetheless others, most notably local public representative, Tony Gregory TD finds the procedure difficult to understand and criticizes its operation in the recent past:

> The process is that nominations from community groups can go in from all over the docklands area and those nominees and details go to the Minister, and the Minister picks from them the ones that he wants. And the best attendee and the most effective in my opinion of the first five years of the docklands council certainly from the northside was Gerry Fay, and because he wasn't anybody's hack – having attended every single meeting, having been involved in everything strictly on a voluntary basis – he got a letter at the end of the five years, thanking him for his time and he didn't believe it. And I remember talking to him at the time telling him 'Gerry you've been stabbed in the back, you'd want to face facts', and he couldn't believe it and didn't believe it until it was clear that other people … were taken on the docklands council. So if that's democracy, maybe it is, maybe it isn't. But I think the essential thing about the Council is that regardless who is on it, it doesn't make decisions.
> (Interview with Tony Gregory, 13 July 2004.)

While this may indeed be the case, the key differences between the original Custom House Docks Development Authority and the Dublin Docklands Development Authority has been an institutionalized emphasis and legislative requirement to achieve tangible social regeneration, be that in the spheres of education, housing or other community development initiatives. Rather than depending on a trickle-down effect, the most apparent change since 1997 has been the high level, interventionist approach to social need and the contribution of the local community to solving their own issues. The appointment of Paul Maloney, who previously held the position of executive manager for Dublin City Council in Dublin Central (primarily the north inner city), as chief executive is a clear example of this change. With a background in the implementation and delivery of social, economic and enterprise strategies to complement major physical development projects, he is ideally placed to understand the challenges, identify realistic targets and foster better relations between the docklands authority and the city council.

Indeed this has been further developed through the reactivation and refocused operation of the Community Liaison Committee (CLC). The inability of the CLC, originally established in 1995 under the Custom House Docks Authority, to wield any influence over the development trajectory of the area resulted in significant scepticism building within the community. Nonetheless, as in many other cities in response to a more collaborative or consensus-building approach to decision-making being adopted, this situation has been turned around in the last five years. The present chair of the Community Liaison Committee, Dónall Curtin, has acknowledged that while it 'was initially viewed with a lot of suspicion … it actually engineered itself very successfully and I suppose that over the course of the four and a half years that I have been involved, we've never had a single vote, which is quite a record'. Commenting on the contemporary situation within the docklands, Ruairí Quinn, TD, also recognizes the importance of this body by describing 'the relationship of the local communities to the DDDA and its personnel [as] very positive'.

The primary role of the Community Liaison Council within docklands has been to develop community capacity to respond to local needs. Appointed by Minister Noel Dempsey in 2002, the seven community representatives are drawn from communities across docklands and would generally be perceived as having significant standing within their own communities. The development of a more participative and transparent approach to planning and developing docklands has been facilitated through this committee, which complements the docklands Council, and meets monthly to 'drill down into the key issues' (Dónall Curtin, CLC chair) and develop projects in conjunction with key institutions and agencies within docklands. The key role has been to develop the social infrastructure of the area as well as addressing the two critical issues of housing and education. The cross-sectoral connections forged by the activities of the community representatives who work at the coalface and can clearly identify the most pressing local issues has been a key to the success of social activities. For example, one of the community representatives, Geraldine O'Driscoll is an active member of the North Wall Women's Centre that has a crèche facility available to support young parents who want to return to the workforce. Similar to the manner in which early regeneration in docklands was designed to pump-prime or encourage physical regeneration through property incentives, one of the recommendations of the CLC was to support the North Wall Women's Centre through subsidies and grant aid for a particular project. Drawing on the financial resources of the DDDA and the project idea of the community leaders, a back to work scheme – the Young Mother's Self Development Programme – was put in place to bring the young women up to a certain level of education, and train them in presentation skills, such as how to prepare a curriculum vitae, how to dress for job interviews and similar techniques. The DDDA engaged the National College of Ireland and Carr Communications to deliver the project and Dónall Curtin argues that it:

> had a massive power, a knock on effect, within a community like that because others said I want to do that next year, I can do that also. That would be impossible if the North Wall Women's Centre did not have a crèche facility. With the best will in the world, with NCI and Carr Communications being available, without having that creche facility, those young women didn't have a chance of going back into the workforce. So it is social infrastructure like that that we will fight for.

Although there have been major strides in achieving improved social facilities and addressing long-term need, others would argue that the DDDA still do not listen sufficiently to the community voice. Within docklands, there is what might be termed a very fragmented institutional landscape in terms of community representation. At least twelve community residents groups represent different interests within the area. It is no surprise then that there is a diversity of opinion on the success of the DDDA in delivering on their remit. Some of those not directly involved with the Community Liaison Committee still argue that:

> they're developer driven rather than community driven … the other bits that go in just cover the cracks in case you think it's totally developer-led … but what they should do is back anything that comes from the community. What the community needs is for the authority to back them.
> (Mairéad McGrath, East Wall Community Development Council.)

The contrast between this perspective and the story told by members of the CLC raises a question regarding the representativeness of this committee. Yet the CLC members would point to the annual docklands social regeneration conference, which has been held in Killarney for the past four years, as an example of its democratic credentials. This conference is an open forum and opportunity for anyone to participate, debate and raise issues and highlight problems of representation if they exist. It is clear that this has not occurred and there appears to be a wide representation from a variety of different interest groups at each conference. There is no evidence that those who may be disillusioned with the process have boycotted the conference.

The disagreement between groups regarding the efficacy of the authority is probably less to do with problems of representation and more about the lack of engagement and transfer of knowledge between the diverse communities within docklands. Government agencies and reports seem to consider and treat the docklands as a single entity, yet the feeling within the area is very different with a clear sense of distinct community identity both north and south of the river and also within areas such as North Wall, East Wall and Ringsend/Irishtown. While local activist, Mairéad McGrath suggests that the 'DDDA say the communities are all one', Dónall Curtin (chair of the CLC and authority board member) sees the situation quite differently. He argues that 'of course they are all different, they all have their own sense of history and culture. But look at Dublin. If someone asked if Rathmines and Ranelagh are different ... of course they're different but does that mean that someone from Rathmines can't go into Ranelagh and have a pint and that sort of thing?' He believes that, through the annual conference organized by the docklands authority, one of the clearest and most useful outcomes has been the fact that different community groups who never interacted with one another are now building up friendships, sharing ideas and together promoting active citizenship. One of the key ways through which the local community will become increasingly empowered in the coming decades is through access to education and life-long learning opportunities.

Education in docklands

> There are 2,000 people in this area. Most of them were born and reared here, their fathers worked on the docks. We had schools, shops, a community centre, all the facilities. The people are 'bang on'. Most of them are dockers sons. There's 70 per cent unemployment, but it's not that we haven't the brains – we haven't any opportunities.
>
> (Gerry Fay, *Irish Times*, 14 July 1987.)

Following the decline in manual occupations and the migration of low skilled economic activities out of city centre locations in the decades after 1950, the areas that suffered most acutely were those where low levels of educational attainment predominated. Docklands was one such district and any real

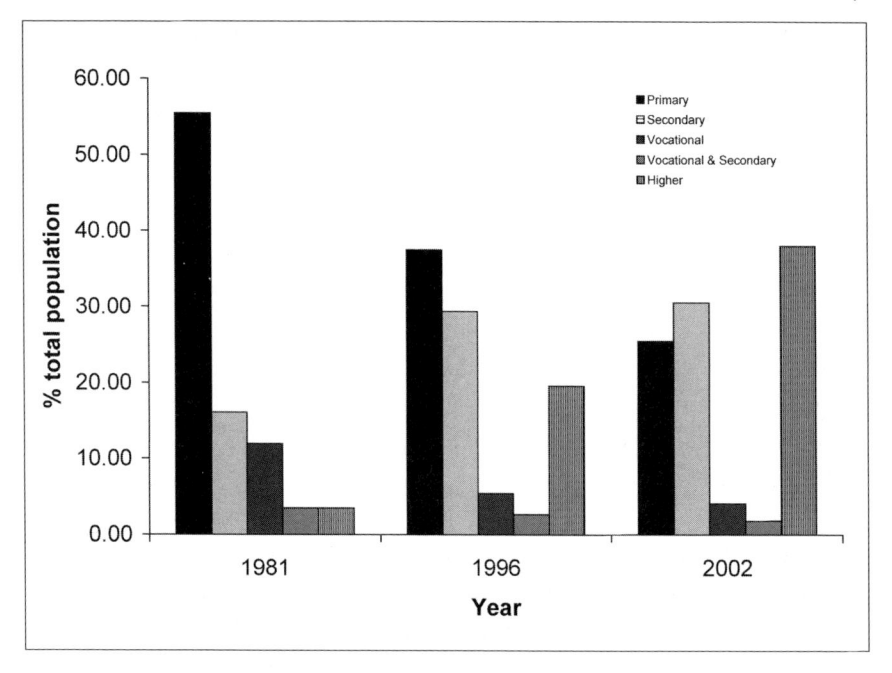

107 Changing levels of highest educational attainment within docklands.
(Compiled from Census of Population of Ireland, various years.)

attempt at meaningful social regeneration had to tackle educational problems as a pathway to addressing broader disadvantage. In the early years of redevelopment at the Custom House Docks, educational programmes played little or no role in the overall vision for the area. The poor spin-off benefits for local residents from the financial services activities attracted to the area can in part be explained by a lack of re-training opportunities. Thus in the development of a new Master Plan for the wider docklands district in 1997, education became a key priority for local residents and has been recognized by other stakeholders such as Dónall Curtin, who has acknowledged that 'there is still a lot more to be done, [but] education is the one thing that kills social disadvantage'.

A measure of the major transformation that is beginning to take effect are the improved levels of educational attainment that have become apparent in the last decade. There has been a steady increase in the number of students (those engaged in full-time education) in the area from 7 per cent in 1986 to almost 14 per cent of total population aged fifteen or over in 2002 (Figure 107).

Further evidence highlighting the beneficial impact of recent regeneration programmes includes the results of the ESRI survey of education levels among young residents comparing the situations in 1997 and 2005 (Table 8). Even

Table 8 Comparative levels of educational attainment within docklands.

1997	2005
35% of children drop out of school before the age of 12	13.3% of children drop out of school before the age of 12
65% before the age of 15	30% before the age of 15
10% sit a school leaving certificate	59.7% sit a school leaving certificate
1% further education	10% further education

(DDDA Social Regeneration Presentation 2006.)

though major improvements have taken place among local residents, the levels of participation at higher levels are still well below the averages for the city as a whole and even for docklands in general, where the number of people having completed third level education has doubled from 19 per cent to 38 per cent. The high proportion of graduates now recorded in the area is probably linked more to the influx of new residents, many of whom are occupying rental accommodation in the new apartment complexes. Local representative Tony Gregory agrees with this analysis stating that:

> I'm sure those figures are the new population coming in because they are very distorted. I don't think there has been any significant change at all for local young people. I mean there would be a small number because of the changes in any case that have gone through to 3rd level, but there is nothing of any significance that I am aware of.

The overall picture in docklands remains one of extremes, with the highly educated new residents living side by side with the longer-term residents for whom education is still a major challenge.

In recognition of the entrenched nature of educational disadvantage, the docklands authority recognized, almost from the beginning of their operation, that a focus on improving life-long learning would have to be central to any plans to reach the targets identified for social regeneration. The concept of *Saolscoil*, literally meaning life-long learning, was a core objective of the 1997 Master Plan. It aspired to the 'promotion of increased access to education and training for all residents in the Area' (DDDA, 1997, p. 24). The key attributes of *Saolscoil* were and are that education is the key to social and economic regeneration, that the ladder of opportunity should be afforded to all who wish to access a higher educational standard, and that intervention by state agencies is required to meet the challenge of raising education and training standards. Although the Master Plan did not specifically aspire to creating a new Third level institution in the area, it did intend implementing a range of educational services for local residents and for the business community, and developing a Centre for Educational Access and Community Development. The DDDA engaged in extensive consultation with a range of third-level institutions from around the city to chart the best way forward and initiated a tendering process for service provision. Trinity College agreed to provide a range of outreach services as part of the pre-existing TAP (Trinity Access Programme) and forged close links with St Andrew's Resource Centre based on Pearse Street. But the extent of 'catch up' that was required to create a level playing field in terms of educational opportunity and achieve the objectives identified in the Master Plan required a more radical approach.

Docklands: a campus without walls

In October 1998, the DDDA identified the National College of Ireland (NCI), then based in Ranelagh (south Dublin), as a preferred educational partner. The National College of Ireland was formerly known as the National College of Industrial Relations and specialized in distance learning education. In contrast to the mainstream third level institutes at the time, many of the degrees offered by the National College of Ireland were part-time, catering for the large majority of students who were also full-time workers. Joyce O'Connor, former President of the NCI, believes that the concerns about the feasibility of the partnership that the Department of Education displayed at the time were primarily due to misapprehension of both their student numbers and the costings for the project. She argues that there was

> a misunderstanding that we were dealing with full-time students. It was a different model. We didn't conform to any of the norms. We weren't a State university. We didn't have the traditional State population of 17–21 year olds … we had those but we also had this other group. So when we were proposing to move, I suppose there was an issue that we weren't State and secondly, I think they thought it was 8,000 full-time students [we were envisaging] … it was up to us to communicate with them what we were doing, so we've a really good working relationship with them now. We are different and that's the

way we want to stay. We can be innovative, we can do things, we're very flexible. ... Money was never the issue because the State only really contributed less than 10 per cent because we engaged the business community, the unions, the community and at the end of the day it was all those stakeholders who could see that what we were doing was what was needed for this particular area.

In many ways it should not be surprising that the National College of Ireland (NCI) was identified as the most appropriate institution to deliver on the educational objectives of the Master Plan. It was no secret that the college had outgrown the existing campus at Ranelagh and was looking for a new home. It is possible that the decision to locate to docklands may have been influenced by the fact that Sean Fitzpatrick, the brother of Joyce O'Connor, is a Director of Anglo-Irish Bank based in the Custom House Docks and has been a member of the Executive Board of the Dublin Docklands Development Authority since the late 1990s. These personal connections may have provided the decision-makers in the NCI with a very clear picture of how the docklands project was likely to evolve and their potential role within it.

From the early 1990s the NCI had been developing a very specific mission: improving access to education and acting as a bridge between business and the community. Given the educational needs of the docklands area and the location of the International Financial Services Centre, a docklands location could not have been any better. O'Connor acknowledges that this kind of mission within third level institutions has become very ordinary in recent years, however she also suggests that:

> it wasn't at the time ... and in a way what *Saolscoil* was, involving preschool, parents, bringing together people in the community, was actually what we did. So we were doing it already. There was a synergy between what they [the DDDA] wanted to do and what we were doing and the other thing is that we had been working down here for about three to four years before we moved down. So we had built up a relationship with the local community.

The theme that runs through the overall vision for educational improvement in the docklands is accessibility and this is reflected in the concept of the NCI as a 'Campus without Walls', not just limited to a city-centre campus but including outreach programmes, business training and online learning for students throughout the country. This idea fits well with the special challenges that a docklands location presented and to support the development, the DDDA provided a 0.5 hectare site to the NCI free of charge. Originally destined for Spencer Dock, the planning problems that were emerging in that location resulted in a new site being identified for the college on Mayor Street. Developed at a cost of €31 million and opened in September 2002, the campus was funded by both the sale of lands in Ranelagh but also by corporate fundraising reflecting the idea of its community base as both the business and the local community. Among the biggest supporters were Citibank (Dublin) and the Citigroup Fund (New York) who contributed a total of £135,000 towards construction of the building. One of the biggest players in the property industry in Dublin, Patrick Kelly, who was the developer of the project, also contributed €250,000 towards the construction costs. Kelly has been involved with many major projects throughout the city including Smithfield and has a close business partnership with Anglo-Irish Bank. Kelly's involvement in docklands extended to the development of the Clarion Hotel, adjacent to the NCI, and also to the construction and management of the student residences associated with the college. According to a recent report:

> Kelly received tax credits for the student housing, too. Under an agreement with the college, Kelly's team will own the apartments for a decade and reap tax credits. In 2012, the college will buy the dorms back for $26 million.
> (*Southwest Florida Herald Tribune*, 7 April 2007.)

As well as the docklands authority and the construction industry, local residents also played a key role in the design of the building which fronts on to a new public open square at Mayor Street. It is designed as an open book to reflect the idea that the college is a central part of the local community (Figure 108). This has been facilitated through the decision to open facilities such as the college restaurant and gym to the public as well as to staff and students. An early learning centre has been provided to fulfil the pre-school element of *Saolscoil*, adjacent to new retail facilities and student residences. The former President, Joyce O'Connor, argues that:

> We see ourselves as living in the community and providing literally education from the cradle to the grave. The building really is a physical representation of the mission of the college. It's open, we're open seven

108 National College of Ireland. (N. Moore.)

days a week for courses ... Everybody is involved in learning, so creating that learning environment starts as you come in the door. It's not just about the faculty. So in other words what we talked about in our mission was a reality.

Partnership and inclusion is a key part of the strategy being developed to combat educational disadvantage and this broadens the way the institution operates beyond just traditional educational delivery. As part of the original agreement with the DDDA, there is a quota of places reserved for inner-city residents and a further quota reserved for dockland residents. Students who reside in disadvantaged areas throughout the country can apply for a place in the college through what is termed the 'Area-based partnership' route. Additional programmes to attempt to reverse entrenched negative attitudes to education have been developed with the North Wall Women's Centre, with the various community groups in mentoring, and with the local primary and secondary schools. In some cases and reflective of the outreach ethos, the programmes often start outside and are brought in to the college at a later stage.

The types of programmes that have been developed go beyond those aimed at improving access to higher education because many of the real educational hurdles occur much earlier. In order to combat these critical needs, pre-school programmes have been identified as a major concern for the near future. In conjunction with Yale University, the NCI has been examining best practice in supporting pre-school education and has created a Centre for Childcare and Adolescent Studies. The overall aim will be to promote the notion that the community itself is a learning environment and existing community facilities such as the North Wall Women's Centre are being encouraged to become early learning centres. At primary and secondary level, the key issue remains how to encourage students to stay in school. Although there is a major improvement in retention levels, 30 per cent of students still drop out of school before the age of 15, reducing their chances of employment in later years. A critical need to educate parents about the importance of their children staying in school and the need to provide a supportive environment for them has resulted in the establishment of the Parents in Education Initiative. Since 1998, 158 local parents have attended the course and today one of those parents is working as a facilitator to, and mentor for, other parents. The Centre for Educational Opportunity within the NCI, also fulfils the remit of the proposed Centre for Educational Access and Community Development in the 1997 Master Plan, by providing an access point for community information and integrating the work of the college with the community. A range of other schemes and initiatives have also been adopted in a bid to reshape the educational profile of long-term residents, including the Discovering University programme which targets third, fourth and fifth year students at secondary school. The aim is to offer students an opportunity to set their sights on a college place early in their secondary school career and give them first hand experience of different career choices available to them. From its inception, 230 students have partaken in the week-long NCI programme where they are introduced to study skills, personal development skills, group decision making, group problem solving and enterprise skills. This kind of programme is designed to encourage students to aspire to third level participation and is complemented by a range of after-school study programmes for Junior and Leaving Certificate students, as well as a range of qualifications in community development.

While these are worthwhile programmes, there has also been a need to address more basic requirements. One of the most interesting patterns that has emerged in recent times is the trend towards educating local children outside the area. This has a detrimental impact not only on the continued survival of educational facilities within the area but also in terms of developing the school as a key focus of community activity. There are many reasons for it, one being that children have less chance of being socialized into 'dropping out' or engaging in anti-social behaviour if they are schooled outside the local area where these problems exist. This is not particular to docklands as similar patterns are evident in other disadvantaged areas throughout the north and south city. Local representative, Tony Gregory, TD, has also identified a more basic reason for the emergence of this trend. He believes that many parents fear for the safety of their children in local school buildings which were starved of resources for many years:

> For a long time, the only local school in the North Wall area at primary level was a building that was literally falling down and I used to go into the docklands people and they'd be briefing us on some nice new schemes and scholarship things, which are all very well in themselves and welcome and all the rest if we want to equip these kids for the jobs that are coming down the line and educate them, and I just pointed out that up the road because of a school building, parents won't send their children into the school. And now that was about two years ago, and after a long campaign by at least three different principals who went through it, that was refurbished. But I mean that's the level of it, there is nothing else by way of planning the educational needs of kids down there. You know the school in East Wall is in the same boat, very run down, under-resourced, on the edge of the main road into the port and so on.

He has identified the redevelopment of a new primary school at Seville Place as one of the emerging controversies in the area. Given the state of disrepair that the buildings have fallen into, a proposal has been forwarded to refurbish and extend the school by creating a 5-storey building on the same site close to the church.

But there is disagreement as to the best way forward with the DDDA and some community groups arguing for this higher building on the existing site, and others demanding that a new site be set aside for the development of a modern school. Tony Gregory does not see

> how it would possibly work but rather than giving a site which would be worth a lot of money for a modern school, the docklands are jumping at this idea of building five storeys up around the church which I think is mad myself. I can't see it happening but some of the local groups feel that if you develop a new school on a new site, you

will destroy the village or community base around north wall, around Seville Place, destroy the little schools they have by dragging everybody down to this new building.

Yet Dónall Curtin disagrees with this analysis, citing the development of the new School as an example of the commitment of the docklands authority to the local community. He describes the view of the Board of the DDDA as a determination to 'go with real architectural excellence … to create a wow factor within the community rather than going with something functional, that fits the budget … within the community itself people will be proud and say this is our school and it will build self-confidence'. This difference of opinion and lack of consensus reflects a much broader debate and divide within docklands as to the success or otherwise of the current approach to regeneration. While Gregory has argued that the above example shows a lack of coherent planning in delivering the educational objectives in the docklands Master Plan, Joyce O'Connor disagrees, arguing that all the programmes now in place are being targeted in a strategic fashion to deliver over the long-term for the traditional community. Those who view the educational transformations as perhaps the biggest achievements in social regeneration, point to the activities now being provided for students at a range of levels as well as the development of links between the educational and business domains in the extended IFSC area, as clear examples of success and a vindication of the current approach being taken.

An indicator of the transformation that has occurred, as well as the uniqueness of regeneration, in docklands is measured by the manner in which a range of State and non-statutory agencies have cooperated to adopt an integrated approach to development within this area. As well as setting a range of objectives for local social and educational improvements, a key aspect of the activities of the DDDA is to support the continued development of the International Financial Services Centre and develop an environment conducive to the retention of financial services operations even after the tax breaks cease. The Industrial Development Authority, Irish Financial Services Regulatory Authority, Central Bank and the Department of the Taoiseach, have worked with the National College of Ireland to set up the International Financial Services Institute intended to provide the upskilling and training programmes needed to support the needs of financial services. While this is not only targeted at companies within the IFSC, but throughout the country, the key local benefit has been the development of the Pathways to Employment

109 Dock Mill apartments, January, 2005. (J. Brady.)

programme. The various agencies have worked together to attempt to identify the needs of business and how the local community might meet their requirements. This may seem a surprisingly altruistic activity but the key reason two major financial players, Bank of Ireland and ICS Building Society, initially became involved was to resolve their difficulties of staff recruitment and retention. Pathways to Employment guarantees local people who successfully complete a training programme with NCI, a job in the financial services sector. The power of this programme to break down real and perceived barriers to employment is critical. Traditionally, the closest involvement that many within the local community could have hoped to have with the financial services operations would have been working as low paid cleaners or security personnel, with little opportunities for progression and diversification. Other programmes developed in conjunction with the business community to aid student retention include the job placement programme that guarantees students who remain at school and complete their Leaving Certificate a job in the IFSC. Twelve different companies have been engaged in the scheme run by FÁS and to date 120 local young people have participated successfully. A participant, Liam Colvin, who was placed with Bank of Ireland Outsourcing Services Ltd described the opportunity as 'a great experience, helping me further my education, and make decisions for my future' (DDDA, 1999).

In conjunction with Reuters International News Agency, the Dublin Docklands Development Authority also established a Third-level Scholarship and sponsored the attendance of local students at university. The first academic year of this award was 1998–9 and between then and 2005, 160 students benefited with only eight students failing to complete their chosen programmes. This kind of intervention at the local level is unique. Many similar projects that have been undertaken in other cities globally have failed to deliver any connections between major economic players and the local community. Direct intervention on the part of the State and this partnership approach is rare, but is perhaps becoming more common as major corporations are increasingly held to account and are criticized for lack of social responsibility. Being seen to engage in this kind of activity provides significant spin-off benefits and opportunities for publicists in return for what is really a very small investment in corporate terms. While it does seem to appear that the view of the DDDA (2002) that 'education is the bedrock of sustainability in the regeneration process' is informing the entire programme, the viability of long-term docklands communities within a transformed sector of the city will depend on appropriate solutions to the most contentious issue in the area: access to affordable, quality housing for local people.

Accommodating development

In common with the situation in many urban centres around the country in recent years, housing has become a critical issue and a bone of contention within docklands. The same factors of inadequate housing supply and lack of affordability that have forced those who work in the city to commute anything up to 80km to and from work each day, compounded the already thorny issue of housing in this district. In the early 1980s, the problems were characterized by the lack of private sector interest in constructing dwellings within the area and the financial inability of the local authority to engage in anything but small-scale projects. Now, it is impossible for developers to keep up with the demand for housing in this particular area, a demand that was nowhere close to being sated by the completion of 1,852 dwellings in the area in 2005. The most recent census would suggest that population growth in this area is on course to meet the target of a 25,000 person increase between 1997 and 2012, a radical turnaround from the situation in the late 1980s when regeneration was first proposed for docklands. Most of this growth has been accommodated in apartment developments, aimed primarily at singles and couples rather than families and so it may be reasonable to suggest that the manner in which physical redevelopment has taken place has directly led to the emergence of a very specific social structure in docklands as indicated earlier in the chapter. The survey undertaken by Hooke and MacDonald for the DDDA in 2005 discovered that 60 per cent of those who had purchased an apartment in the previous three years were single and that there were few family units. Households with children accounted for 11 per cent of those already living in docklands and only 7 per cent of those that had recently purchased.

This trend first became apparent during the initial phase of redevelopment in docklands when residential construction around the Inner Dock was entirely apartment-based. The 1997 Master Plan for docklands had specifically aspired to promoting a more balanced family and social profile within the overall area through the creation of different kinds of housing, yet the emphasis on apartment dwellings has remained strong. In ways this is more to do with a general cultural shift in the city; 90 per cent of new homes constructed during 2003 across the Greater Dublin Area were apartment dwellings as opposed to the traditional semi-detached suburban home. This looks set to continue and the trend is clearly evident in the new developments coming on stream both north and south of the river. These include the Dock Mill and the Gasworks at Barrow Street, Crosbie's Yard at Ossory Road and Gallery Quay at the Grand Canal Dock, all apartment-style developments with a limited number of own-door duplexes (Figure III). Many are quite large and some of the three-bed apartments are as large as a small townhouse. Nonetheless, the attractiveness and affordability of apartment-style living to young families is questionable. The cost of a two-bed apartment launched at Dock Mill in 2004 began at €370,000 with an additional €35,000 price tag on a car parking space, while at Gallery Quay the starting price range for a similar apartment was €435,000. Recognition of this problem of affordability is reflected in the 2003 Master Plan (Policy 4.2.6.10) that states that 'the authority will promote conditions and incentives that will render docklands an attractive location in which to start and/or raise a family'. While images of parents standing on a balcony overlooking the waterfront holding their babies is a public relations dream, the reality of how appropriate this type of housing design might be to young families is unconsidered. This is not unusual as city promotion throughout the world has become an intensive PR exercise. The evolution of the 'right image' to attract 'the right sort of people' has been of paramount importance and understanding why this 're-imaging' occurs is the key to comprehending the contemporary dynamics of city development and management. Like any other commodity, the city is being actively marketed and

110 Clarion Quay, completed development, 2005. (N. Moore.)

111 Gallery Quay first phase nears completion, 2004. (J. Brady.)

sold as areas like docklands fight off their historical image of industry and de-industrialization. In order to be successful they must adopt a post-industrial image highlighting community parks, security, accessibility to a range of functions and participation in particular kinds of lifestyle as major selling points.

In many developments, not just in docklands, the marketing blurb misleads the reader on the geography of the city with proximity to particular amenities being exaggerated. Thus the Gallery Quay development promotes 'the excellent dockside views, with clean, uncluttered modern lines' and 'access to the central private landscaped courtyard'. Since their construction, it has become quite apparent that many of the apartments in this development have no visual connection with the riverfront at all. The promotional material for the Dock Mill apartments at Grand Canal Dock focuses on accessibility, a major advantage of the development being that it is 'superbly located, with the benefit of a Dublin 4 address, next door to the new DART station on Barrow Street, within walking distance of all the amenities of Ballsbridge and the South City'. The kind of lifestyle being promoted through marketing material at this development is highly specific and aimed at traditional apartment dwellers, young twenty-somethings or couples with no children. There is

> a variety of restaurants, cafés, bars, hotels, shops, libraries, sports clubs and leisure facilities, Lansdowne Road Rugby Stadium, the RDS and Herbert Park. There is convenient access to the IFSC, Trinity College, Temple Bar and all the hustle and bustle of the nightlife and entertainment venues of the city.

There is no mention of schools, playgroups or crèches, elderly day care centres or social clubs. Although the Dublin Docklands Development Authority, its predecessor and other agencies have had huge success in revitalizing the population of the city centre, there is a danger that promotional activities targeted at very defined populations will create an unbalanced population structure and alienate as many people as it attracts.

One of the major disappointments for those who held such high hopes for regeneration is that this new style of living has failed to support a community spirit, traditionally very strong within the inner city. Management consultants, KPMG have suggested that where the cost of units has been higher, the proportion of owner-occupiers is lower and corporate investors become more apparent (Department of the Environment, 1996). One newspaper article recently reported that when the Sheriff Street flats site was rebuilt as Custom

House Square, Chesterbridge Ltd retained 33 apartments for investment purposes, so confident were they of the future prospects for the area (*Sunday Tribune*, 9 May 2004). When the remaining 47 apartments were released for sale in November 2000, 75 per cent were purchased by investors. During the first decade of redevelopment, private rental became the dominant tenure in docklands in contrast to the strong tradition of home ownership in the city. In more established residential districts within docklands, 66 per cent of dwellings are owner-occupied whereas at developments, such as at the Custom House Docks, the balance of tenure is 20 per cent owner-occupation and 80 per cent rental. Early reports suggested that in a number of cases, investors purchased more than one apartment and often these were well-known personalities on the Irish social scene. Entrepreneurs who invested in the area included restaurateur Patrick Guilbaud and his wife, who purchased two apartments, and Liavan Mallin, the co-founder of Celtic Hampers who owned four properties at the time of interview (*Sunday Tribune*, 9 May 2004). In docklands overall, a large proportion of total housing units are rental apartments and in Ireland where owner-occupation is an extremely strong tradition, this marks a radical cultural and social change. Many of the new residents are young, foreign nationals, and this has dramatically altered the socio-economic and demographic profile of the inner city. In the overall docklands zone 15 per cent of residents are non-nationals; in the recently regenerated districts such as the Custom House Docks this increases to 28 per cent and at South Dock to 24 per cent based on the most recent detailed census figures available. In such circumstances where people have less in common and have less allegiance to the place, community spirit is difficult to build and maintain. The transient nature of the population became a significant cause for concern within the earliest developments, with one resident commenting that:

> British and German apartment dwellers go home at weekends. There are always a lot of new faces around, because you get people working on six month and one-year contracts.
> (*Irish Independent*, 3 February 1998.)

A survey of apartment residents undertaken by KPMG in 1995 for the Department of the Environment supported this view of the apartments, particularly those closest to the Financial Services Centre. Their research concluded that market forces and the design of residential units were the primary determinants in encouraging transience. The size of apartments was not conducive to encouraging people to make their permanent homes here and spiralling property values encouraged residents to speculate and make quick profits. Of those interviewed, 25 per cent stated that they would move within two years, while a further 55 per cent did not intend residing at the same address for more than five years. Only 20 per cent of residents in the Custom House Docks complex in 1996 envisaged themselves living there for longer than five years (Department of the Environment, 1996a). This trend reinforced the distinctions and the commitment to community that set many private apartment dwellers apart from those residents in local authority and affordable housing. Brian O'Gorman of Clúid Housing Association summed up how this difference manifests itself today by suggesting that public sector tenants

> will be putting down roots in Spencer Dock and a lot of small businesses and schools will benefit over the long term as a result. This is in contrast to the private apartments which are more likely to be either rented out by investors or owned by young professional singles on the move and will have a more transient population.
> (*Irish Times*, 4 March 2003.)

For many people used to living in neighbourhoods with their own front door and garden, tiny apartments were understandably viewed as an unattractive accommodation option for those with longer-term plans. The problems that this created for the inner city in terms of both long-term sustainability and the immediate delivery of social infrastructure resulted in State intervention in 1995. In an attempt to increase the attractiveness of this new mode of living, a lifestyle norm in most European cities, the Department of the Environment issued a set of guidelines in 1995 aimed at ensuring a wider mix of size and types of accommodation in designated areas. Recommendations included guidelines covering the size of apartments, required amenity facilities, recommended internal design and the numbers of apartments permitted per lift in a development. As these guidelines were issued after the construction of units in the original Custom House Docks development, this complex is one of the few developments within docklands that fails to comply with the minimum requirements. Yet the tendency of many developers to just barely meet minimum unit size requirements combined with the high prices within docklands did little to discourage investors and encourage owner-occupiers until the recent past. As late as 2003, the docklands authority concluded that 'to date market housing has been predominantly occupied by short to medium term residents without children, who demonstrate a lower level of commitment to the Area' (DDDA, 2003, p. 40). Confident in the buoyant nature of the market and emboldened by the high demand for development sites within docklands, the DDDA attempted to intervene by introducing new guidelines indicating the percentage of each unit type that must be provided in every development for which they grant permission to develop (Table 9). The percentage of units in each category completed in 2005 suggests that such a prescriptive approach has begun to succeed in its objective. The recent survey undertaken by Hooke and MacDonald of new owners within docklands also reinforces the benefits from such an approach. They estimate that 62 per cent of those who purchased in new schemes developed between 2002 and 2005 were owner-occupiers with investors only accounting for 38 per cent of new buyers. This increasing trend towards owner-occupation bodes well for the future sustainability of docklands as a vibrant urban district. What is undoubted is that without the provision of shared, family facilities to encourage social interaction, levels of distrust will remain high between the older and newer residents, the former of whom have become sceptical about the purpose of redevelopment while the latter have become increasingly entrenched in high security enclaves.

Table 9 Proportion of unit types in new apartment complexes in docklands.

Accommodation size	Desired % of total units	Completions in 2005
1 bedroom	25–35%	22%
2 bedroom	35–45%	58%
3 bedroom	20–25%	20%
4 bedroom	0–5%	0%

(DDDA, 2003, p. 44 and DDDA Master Plan Monitoring Report, 2005, p. 24.)

Housing tenure is not the only challenge related to housing in docklands. Accessibility to quality housing remains a key problem as economic regeneration has altered the general perception of docklands rendering it one of the most desirable locations to live in Dublin. Although this is economically beneficial, the development of gated communities in high-cost complexes has reinforced pre-existing social divides. As we already discussed in this chapter, many new affluent residents live behind security fences, engage in minimal interaction in the immediate area and have a low commitment to building a sense of place. One of the most significant concerns for the long-term community until relatively recently, in what could be perceived as a reversal of the traditional public-private role, was that these new private developments would become neglected and rundown due to sub-letting and the prevalence of absentee landlordism. This was also recognized as a potential long-term problem and attempts were made by the Government to guard against this difficulty during the establishment of the DDDA. Section 24(2)(b)(vi) of the Dublin Docklands Development Authority Act proposed that any Master Plan should 'include proposals for the development of existing and new residential communities in the Dublin Docklands Area including the development of housing for people of different social backgrounds'. This was encapsulated in Policy 2 of the 1997 Master Plan which stated that 'the authority will promote the development of new housing which will reflect the diversity of needs in any community, including housing for couples with children, housing for single parent families, sheltered housing and housing for people with disabilities'. By the time this Master Plan was revised in late 2003 achieving this social mix still remained a problem. Partly this was due to the slow pace of residential development in the area, which had only resulted in 1,500 units constructed in the first five years of redevelopment up to 2002. The population has increased by about 2,000 people, but this is well below the target of 5,000 people set down in the 1997 plan. Over the next year with the large developments of Spencer Dock, Forbes Quay, Longboat Quay and other similar developments nearing completion, this will increase significantly (Figure 112).

Although the actual construction of apartments may have been quite slow, one area in which the DDDA has actually led the way has been in the mechanisms to deliver affordable and social housing in an economic climate that is so driven by market growth. To counter the earlier concerns that emerged from experience in the initial schemes close to the Custom House Docks, the local community negotiated a clause in the 1997 Master Plan which stipulated a requirement of 20 per cent social and affordable housing in all new developments. This has been exceeded in the most recent year for which there are data available (2005), with 20.8 per cent of houses delivered under the social and affordable housing schemes. This provision, negotiated by and on behalf of the community, has subsequently become embedded in national planning policy through Part V of the Planning and Development Act, 2000. Dublin City Council's most recent housing strategy has indicated that 10 per cent of all new developments throughout the city should be social housing and that 10 per cent will also be provided at affordable prices. While the introduction of this quota system was a hard-fought battle in the

112 Hanover Quay, late-2004 – Grand Canal Basin. (J. Brady.)

113 Forbes Quay under construction, November 2004. (J. Brady.)

114 Hanover Quay, dockside, with Longboat Quay, October 2005 (J. Brady.)

115 Longboat Quay apartments, July 2007. (J. Brady.)

negotiations over the original Master Plan, it has subsequently become an accepted norm. The Dublin Docklands Development Authority state that the 'key objective of [their] Housing Policy is to allow the existing community to continue to live in docklands and enjoy the social and economic regeneration of the area. We will provide a wide range of new housing in order to achieve a good social mix through the integration of new residential communities with the existing local communities in the area' (DDDA website, 12 July 2004). The recent increases in the pace at which housing is being completed will go a long way towards achieving these objectives, as there is no shortage of people requesting accommodation in docklands. A major change is that much of this new public-sector accommodation will be provided through the private sector. In essence, the pace at which private developers construct housing schemes will control the length of the municipal housing list. A positive trend in terms of housing delivery has been that the downturn in the commercial market around the time of the publication of the Grand Canal Dock Planning Scheme resulted in renewed emphasis on the residential sector. In an attempt to maximize market returns, the ratio of commercial to residential development was switched in favour of residential. The authority welcomed this shift given that 'the residential emphasis will, for the next period, drive the regeneration process and, at the same time, meet critical housing needs for all sectors, including social and affordable' (DDDA, 2002, p. 6).

There has been widespread controversy as to the wisdom of such an approach, but the experience to date has been relatively positive. The first scheme completed in the State under the new private/social/affordable mix was at Clarion Quay in the northern docklands along Mayor Street (Figure 116). Located adjacent to the National College of Ireland, the complex was a joint venture between Pierse Construction Ltd, Alanis and the docklands authority and consisted of 185 units, 37 of which have been allocated to the social/affordable sector. These are clustered in two blocks within the overall development and were allocated to residents in early 2002. The 37 units were transferred to Clúid Housing association under a rental subsidy scheme in preference to employing a high cost private management company. In its earlier incarnation as St Pancras Housing Association (Ireland), Clúid had already been engaged in managing the public housing stock on the northern edge of the former Sheriff Street and so had some experience in the area and an awareness of the key issues. Increasingly cited as a key objective of the docklands authority, the development of a sense of community and 'sense of place' was encouraged through a four-week pre-tenancy course that residents were required to attend.

116 Clarion Quay apartments. (N. Moore.)

Organized by Clúid and facilitated by a community development worker, this course resulted in the emergence of a resident-based rota system for maintaining the communal areas within the social blocks at a minimal cost. In national terms, this development is perhaps unique as it was the only scheme of its type built before the Minister for the Environment, Martin Cullen, amended Part V of the Planning and Development Act, 2000 after coming under significant pressure from lobbying by the construction sector. This amendment has enabled developers to provide a site elsewhere within the local authority area for social housing instead of building it within the wider complex, defeating the purpose of the social mix argument and furthering residential segregation. Kieran Murphy from the housing organization Threshold, has thus described Clarion Quay as a 'museum piece of Irish social housing history' (*Irish Times*, 17 December 2002). Another view would be that Clarion Quay illustrates how radical housing initiatives can in fact operate quite successfully if the will is there to do so. This view was reflected in a report commissioned by the DDDA to ascertain the views of new residents on their homes.

All of those occupying the social/affordable apartments are long-term inner-city dwellers, with 64 per cent of residents from docklands or the adjacent east inner city. In 31 of the 37 apartments at least one resident lived in docklands prior to moving to Clarion Quay. In common with the general trends emerging in docklands, a large proportion (54%) of new residents were aged under 30 but in contrast to the situation in the private developments constructed in the last decade, 78 per cent of the residents had children with 42 per cent of those children under the age of five. In the survey, all residents agreed that they were highly satisfied with their new home, with 94 per cent rating the accommodation excellent. The contrast with their previous accommodation and living conditions is stark, illustrating a dramatic improvement in quality of life for those that have been facilitated.

> Before coming we were in a bed-sit. It was damp and heating never worked.
>
> It is a better quality of life here. We all get on well and we have good neighbours. I feel more secure and do not feel threatened or frightened.
>
> It was difficult for children living in overcrowded conditions. The kids have their own space and their own bedrooms. It is great.

What is intriguing is the different perceptions of development that have emerged between outsiders and those who have made their home here. It was initially observed that while the public and private housing does not appear to be all that much physically different from the outside, they are still separated into distinct and separate blocks. While social exclusion in the traditional sense of people being forced out of particular neighbourhoods is not occurring, there is significant polarization at the scale of the individual apartment block. Yet during the survey of Clarion Quay residents, separation was perceived as a positive attribute. Residents of the social housing disliked the idea of being 'peppered' throughout the private blocks and argued that being located together gives them a much better sense of community:

> It is great living in yuppyland. We love the buzz and there is a great atmosphere.

There is also a pervading sense that they have very little in common with those living in the private apartments who are perceived by the more settled residents as 'fly-by-nights'. From a management point of view, Clúid Housing Association maintain that clustering is the optimal approach because if apartments are peppered throughout the scheme, maintenance costs rise and the homes cease to be affordable. The residents themselves also prefer it that way ... if they were mixed in with the private units, many of which will be rented, it could result in a clash of lifestyles.

(*Irish Times*, 4 March 2003.)

While there are many positives, a number of issues considered difficult by the social housing residents have emerged. Among these include the conflict over the communal garden space within the development. The diversity of perceptions of this small space illustrates the highly contentious nature of individual areas within the city or even within a particular development. Private apartment residents consider the communal space an aesthetic amenity, a place to be visually consumed from their balconies, while local children have viewed it as a potential play area leading to conflict. Suggestions by the management company to construct a wall that would prevent the children accessing the garden were greeted with widespread resistance (Hogan, 2006). This problem has been compounded by the visibility of a nearby private crèche, unaffordable to many social-housing residents. This restricted access to childcare facilities and amenities for economic reasons is just the latest example of a long-standing inability to plan service facilities in line with social housing as happened in Dublin in Ballymun and Tallaght in the 1960s and 1970s. Following discussions at the Annual Docklands Social Regeneration Conference in 2006 about the importance of addressing this issue, a Childcare Forum has been established and the docklands authority has commissioned a research report on the necessary requirements. Again this is an issue of national importance and concern but it has been brought into sharp focus in docklands because of the unprecedented proximity of diverse social groupings, one of the objectives of the Master Plan.

While the attainment of social and affordable housing objectives has been achieved at Clarion Quay, other developments have not progressed as quickly or been delivered as smoothly to waiting residents. Ten affordable units have been delivered by Zelda properties at Thorncastle Street as part of a mixed-use scheme but only after significant controversy in summer 2003. Part of the original planning application approved by the DDDA under a Section 25 agreement, which meant that the development became exempt from the normal planning process, was that ten units would be transferred by Zelda to the local authority at a cost of €1.9 million. The city council would then select ten purchasers under the Dublin City Council Shared Ownership Scheme.

But on completion of the units in May 2003, a dispute arose between Zelda Properties and Dublin City Council as the developer increased the transfer price by more than €1 million, having a knock-on effect of €100,000 to each purchaser. For many tenants who had agreed to purchase their own homes at the new development, this inflation in the asking price would have rendered it impossible to move. Lorraine Malone, one local authority tenant living in Markievicz House on Pearse Street, described her reaction to the announcement:

> I sleep with my daughter and her daughter in the one bedroom. It's very cramped and we were really looking forward to moving in.
> (*Irish Times*, 4 June 2003.)

In response the City Council sought advice and initiated legal proceedings, which once picked up by the national media was rapidly resolved with the original agreement honoured. In a public statement, Zelda Properties commented that 'simply to relieve the hardship, the company has agreed to transfer the houses at a cost that represents an actual loss to it. The matter is now closed' (*Irish Times*, 6 June 2003).

In summer 2004 in a much more straightforward transaction, 72 social and affordable units were completed on East Road, on a site provided by the authority, selling at €175,000 for a one-bed apartment and rising to €220,000 for a three-bed unit, substantially below market price. The total number of housing units delivered under the affordable housing scheme up to Sept 2006 was 251. Nonetheless, given that the target number of social and affordable housing for the docklands area is between 1,600 and 2,200 units, this is a poor record so far. A number of other projects including the City Housing Initiative will provide 62 units, and the Grand Canal Dock scheme will provide another 139 units (*Irish Independent*, 19 March 2004). By the end of 2006, the DDDA had anticipated that another 130 social and 90 affordable homes would be ready for occupation signalling the first major roll-out of social and affordable housing under the docklands Master Plan. Yet, the speed of completion will have to increase if targets are to be reached. This will be facilitated in future years through the continued development of both the North Lotts and a section of the Poolbeg Peninsula zoned for residential use. The delivery of these will be closely watched, given the tensions that arose during discussions over the social and affordable housing element of the Spencer Dock development, sandwiched between the two long-term residential communities at East Wall and Sheriff Street/Seville Place.

The key problem that residents groups identified with the original Spencer Dock proposal was the lack of integration of different tenure types within the development. Some 947 units were granted planning permission, of which 21.5 per cent would be social or affordable housing thus exceeding the statutory minimum, but locals argued that the urban design would result in 'ghettoization'. The plan proposed developing 600 luxury apartments at the western end of the site, separating it by a main rail interconnector from the bulk of the public and affordable housing, literally on the other side of the tracks. Gerry Fay, Chairman of the North Wall Community Association, argued that 'this is not in the interest of the community. We are going back to the segregated housing and the Sheriff Street legacy. This was the opportunity to move on. I regard the decision as an act of vandalism. We have got to move away from the building of urban slums. This will come back to haunt us' (*Irish Times*, 18 January 2003). While the residents surveyed in Clarion Quay liked the fact that they were clustered together, their block was located immediately adjacent to the privately owned housing, and all units were developed concurrently. If the original plan for Spencer Dock had been proposed, major differences would have been apparent in the location, amenities and speed of delivery of the social and affordable units.

The proposal had previously been opposed by the local community and representatives on the docklands Council for this very reason but had been over-ruled, indicative to some of the lack of democratic accountability of the docklands authority. Tony Gregory, a member of that Council, recalls:

> When the second attempt to develop Spencer Dock was underway the local representatives that were on the docklands council and the community reps and elected reps were unanimous [that they would not become ghettoized] and to a large degree the Council itself supported them. The decision of the Council was for an integration of the social/affordable housing throughout the site, not dividing one half into a luxury development and concentrating the lot in a corner. And that was discussed over about four docklands council meetings, very long, very good debates and at the end of it the Council made a decision, and following that decision and that whole process the authority just ignored it.

Eventually a compromise was reached between the community groups and the docklands authority, yet this highlights how difficult merging the interests of a range of groups and delivering on the housing issue actually is. Dónall

Curtin, chair of the Community Liaison Council, points to the resolution of this issue as one of the key achievements of the CLC. He describes what he saw as the critical issues:

> In the Spencer Dock area there was a discussion about whether it [social and affordable housing] should be in Block M and N or should it be in the luxury developments bang in the middle. But then you have the problems from the Department of the Environment of what they would regard as wasting money for buying a luxury block as distinct to an ordinary block ... For an issue like the Spencer Dock thing, it requires going through all the issues, getting in the relevant people ... People in Dublin City Council would have come in and said how they wanted people grouped together so they could restrict the cost of management charges and that sort of thing arising and, over a period of time, you can workshop through the issues.

A Housing Forum has been established within the docklands to facilitate the pro-active engagement of local communities in influencing the future of their areas. However, it seems that the critical issue over the next few years will focus on housing management and the best model to adopt. There are a number of different methods that have been tried in other areas, and the docklands authority has investigated practice in a range of cities including London, Glasgow and Boston. It would appear that the model of community-led housing in areas like Coin Street in London may be part of the future in Dublin docklands. Locally-based housing management associations may emerge, but the practicalities and details in terms of how this might be delivered have yet to be agreed.

A model for other cities?

Although it seemed unlikely when the docklands project began in 1997, the experience of social regeneration in Dublin docklands may be beginning to provide an international model of good practice for other cities. The mistakes regarding lack of consultation and the primacy afforded to economic and physical development during the tenure of the Custom House Docks Development Authority appear to have been generally avoided in the last decade. There will always be disagreement and those that argue that even more could be done for the local community, yet the situation in Dublin docklands is very different to the experience of waterfront regeneration in other cities which has been almost entirely property-driven. Internationally, redevelopment projects have been characterized by lack of consultation and little community gain, with privatization of the waterfront and significant displacement of local residents and small-scale industries as key features. Only in cities with a strong social tradition and where the local authority played a decisive role in waterfront regeneration has anything similar been attempted. A clear example of this is the Dutch city of Rotterdam where the construction of social housing has been at the core of housing policy for decades. In the Netherlands, the Government considers it a duty to provide housing for any citizen that requires it, and particularly to assist lower income groups raise their standard of living. Therefore, the financing of social housing by private investment is not unusual. One Dutch commentator has remarked that in the Netherlands

> social housing has never been the exclusive domain of the Government and local authorities. Particularly in the case of more expensive housing for rent, institutional investors have co-operated closely with local authorities, whereas building contractors and project developers played an active part providing both cheaper subsidized housing either for sale or for rental through housing associations.
>
> (Kohnstamm, 1993, p. 220.)

So, while the situation that is now emerging in Dublin in relation to the mode and type of housing being developed is new in an Irish context, there is some experience of similar kinds of projects being implemented elsewhere. Nonetheless, these are the exception rather than the norm, with the majority of regeneration projects being criticized for perpetuating and accentuating social exclusion. Community benefits have been seen as an add-on to property development in places like Battery Park City (New York) rather than a core objective. One of the most stringent criticisms of redevelopment in London docklands, for example, in the late 1980s, was that existing residents were displaced particularly as local renewal became overtaken by large-scale economic projects with implications for the national economy. Redevelopment in Dublin may initially have appeared to be operating in a similar manner; in recent years the social projects implemented across a range of sectors have succeeded in reversing this trend. Nonetheless, addressing the challenges of social regeneration is only part of the story of docklands. A number of other issues have proved equally contentious in the relatively short history of this project and four of these provide the subject of the next chapter.

The contested city: Docklands in the new millennium

While the Dublin docklands redevelopment has been widely commended, nationally and internationally, for the dramatic transformations that have occurred in a relatively short period of time, the project has encountered a number of major obstacles along the way. These have emerged primarily as a result of the appropriation of the docklands for new urban uses, many of which are at odds with the more traditional perceptions of the district. The significant investment spent on re-branding and re-imaging this part of Dublin City and the new representations of place that are being constructed stand in stark contrast to the reality of life in this district for many people. The relative ease with which some projects such as the National College of Ireland or the second phase of the IFSC were constructed has not characterized the development process as a whole over the last twenty years. Disagreements between competing groups and interests within docklands have been at the root of many delays, and changes of direction and conflict have become embodied within a number of major developments. This politicization of space in docklands and the increasingly contested nature of four major projects form the focus of this chapter. These conflicts have a number of key themes underpinning them including:

- the fraught relationship between heritage and development;
- the optimum role for public participation and the politics of development in areas subject to renewal;
- the benefits and challenges of environmental regeneration;
- the governance structures necessary to address the apparent imbalance between local needs and city-wide or national development priorities.

The development sites within which these issues have become particularly problematic include the Stack A warehouse on North Wall Quay, Spencer Dock, the Grand Canal Dock and the Poolbeg Peninsula.

Only lately opened to the public, indecision has plagued the redevelopment of the important heritage building Stack A at the original Custom House Docks site from its original designation for cultural/amenity uses in 1987.

More recently, a long-running storm that became bound up with political indecision over the development of the National Conference Centre characterized the initial plans for regeneration at Spencer Dock in the north docklands, while the legacy of industrial activity has posed significant difficulties in fostering redevelopment at the Grand Canal Dock in the south docklands. At the time of writing (July 2007) the future of the Poolbeg Peninsula is still uncertain. In the run-up to the general election in June 2007, community groups, politicians and environmentalists were becoming increasingly exercised over this issue. The recent appointment of a Green Party Minister for the Environment, John Gormley, TD, who also happens to be the representative for this area, may bring a new approach to the national debate on waste policy and in particular on the siting of incinerators. Each of these case studies is specific to docklands, yet they are representative of wider ongoing debates in Irish society in relation to growth and development. The indecision that surrounded the regeneration of the historic Stack A warehouse in docklands was in part a function of its historic protected status, and thus development was not considered commercially viable. Although obviously not comparable to nationally significant sites such as Tara or Carrickmines Castle, the same tension that underpinned the controversies at these locations – whether to favour development rights over heritage – became a major obstacle to redevelopment in docklands for more than a decade.

Heritage and development

Given the extraordinary economic growth rates in Ireland from a low of 2.3 per cent in 1993 to a high of 7.4 per cent in 2006, significant investment in a range of infrastructure, particularly housing and motorway construction, has become a key plank of the national economic development strategy. But one thing that is increasingly contested is the relationship between the dual processes of economic development and heritage protection, as exemplified in the ongoing debates surrounding the construction of the M3 motorway in close proximity to Tara, the most important Celtic site in Ireland. Another similar high-profile case was the long-running campaign to prevent the destruction of Carrickmines Castle, a medieval fortification on the route of the M50, Dublin's orbital motorway. The legal case concluded unsuccessfully from the perspective of those interest groups broadly representing a heritage-environmental agenda, and the motorway opened in late June 2005. Beyond the immediate implications of this decision for this particular area, the long drawn-out saga opened a wider debate in Ireland focusing on 'how and for what purpose we measure the value of heritage' (O'Keeffe, 2005, p. 142).

But it is not only at the national scale that the debate surrounding Ireland's heritages has gathered pace and intensity; the same is true at an intra-urban scale. Over the past thirty years, a number of buildings throughout the city of Dublin have been demolished for redevelopment, none more controversial that the development of the Civic Offices over parts of the medieval town at Wood Quay. In the late 1970s, cultural politics was quite evidently at work in the comments of Senator Gemma Hussey during the parliamentary debates on this issue:

> I would like to remind the Government at this point – this Government which makes such an extremely strong case for the preservation of the Irish language – that they should remember that the Wood Quay question seems to many people to be of extreme importance, no less than making a big effort to preserve our language.
> (Seanad Debates, 6 December 1978.)

The issue of heritage-politics therefore is nothing new in an Irish context, but it is the nature of this discourse that has begun to change in tandem with the transformation of Irish society in recent decades. Even though preservation of the built heritage continues to be a central theme, there is an emerging debate surrounding the commodification and privatization of public heritage, particularly in urban areas that have been subject to urban regeneration. In a recent book on Ireland's heritage, McCarthy (2005, p. 10) has suggested that in the contemporary world the battle lines have moved away from the rather narrow definition of 'heritage' that exercized academics and practitioners in previous decades. The biggest challenge now arises from the dichotomy between 'heritage as a cultural resource and heritage as a capitalist item for consumption' (McCarthy, 2005, p. 10). He suggests that commodification is, above all, driven by the heritage industry, but it is worthwhile considering the other agents and sectors involved in this activity. In docklands, this debate has centred on the redevelopment of Stack A, an historic and architecturally significant piece of industrial heritage located within the confines of the original Custom House Docks area.

Stack A: an ingenious construction

From the beginning of discussions in the mid-1980s on urban redevelopment in Dublin, the Custom House Docks project was portrayed as the flagship scheme that would highlight the benefits of an entrepreneurial approach to revitalization. Central to the Master Plan drawn up by the Custom House Docks Development Authority (CHDDA), was the building within the complex known as Stack A (Figure 117). This is a former bonded tobacco warehouse, built of brick, with one of the finest iron roofs in Europe. It was described in 1821 (the year of its opening) as an 'ingenious construction' (Wright, 1821, p. 7) and is a protected structure. But beyond its architectural significance, the building is historically important, given that it was the only building in mid-nineteenth Dublin sufficiently large to hold a banquet in honour of the Irish Crimean War veterans. Murphy (2002), in his treatise on Ireland's role and reaction to this campaign, suggests that the Irish public followed developments with great interest, publicly supporting the troops leaving Ireland and the tone of ballads sung suggests that Irish opinion was in favour of the war. This substantial banquet was widely reported in the contemporary newspapers and there is little doubt that Stack A must have been familiar to many Dubliners. The banquet was well-described in a contemporary publication:

> There were laid 250 hams, 230 legs of mutton, 250 pieces of beef, 500 meat pies, 100 venison pasties, 100 rice puddings, 250 plum puddings weighing one ton and a half, 200 turkeys and 200 geese, 2,000 rolls, 2,500 lbs of bread, 3 tons of potatoes, 8,500 quart bottles and 3,500 pint bottles of port.
>
> (*Freeman's Journal*, 23 October 1856.)

Mindful of the significance of the building and its historical associations, the original plans for renewal in 1987 proposed a range of public uses for Stack A. These included a night-club/bar area, winter garden, and a cultural/ museum attraction to promote vibrancy but also to ensure that the historic fabric would be conserved and accessible to the general public. This was reiterated in the Master Project Agreement signed between the developers and the CHDDA, which suggested that Stack A would become a hub of activity in docklands, setting itself apart from developments in the rest of the city through the exploitation of its maritime history and environment. Yet at the time of

THE CONTESTED CITY

117 George's Dock and the Stacks in use, 1977. (A. Parker.)

118 Stack A in 1999. (N. Moore.)

writing, twenty years on from the original plans, the building has only recently been opened to the public. It is now an Exhibition and Event venue with a small number of upmarket restaurants. Its exclusive nature means that, for many Dubliners, the opportunity to enjoy and marvel at this remarkable testament to nineteenth-century innovation and engineering remains undiscovered. From the original ambitious proposals, the plans for Stack A were reconstituted numerous times, and this historic building, although in theory providing the greatest potential for revitalization, has in practice proved to be one of the most difficult elements of the docklands project to deliver.

One of the greatest delays to conservation work on the building was that the Hardwicke/British Land consortium of developers eyed the building from an entirely commercial perspective and favoured the development of a shopping centre in the historic warehouse. Although they held an exclusive option on Stack A for over a decade, they were not compelled to deliver an amenity and chose instead to focus their energies on the more profitable elements of the overall project, such as commercial and residential development. In a bid to generate even greater commercial returns, they attempted to have the facility re-zoned to retail use but after much debate, this proposal was rejected. In fact since the original Master Plan was published a range of agencies and other groups have suggested myriad potential uses for Stack A, all with a cultural and public access dimension (Table 10).

Some of them have remained on the table almost from the beginning of the redevelopment, while others have since been accommodated in alternative buildings such as the Museum of Modern Art that is now based at the Royal Hospital, Kilmainham. Each proposal was rejected on financial or operational grounds, and a final call for proposals was made by the Dublin Docklands Development Authority (DDDA) in March 2001. Two of the most widely supported proposals that at various times appeared to have a real possibility of being developed were the Interactive Science Centre and the Museum of Dublin. Their lack of success in obtaining space at this location has meant that the city still remains without these amenities that are considered almost standard components in the urban landscape of other capital cities.

A science centre for the city

From 1986, the development of a science centre within Stack A appeared to be a certainty in that it had been explicitly identified in some of the early plans

Table 10 Proposed development options for Stack A, 1986–2001.

Date	Proponent	Project
April 1986	Dept of Environment	Commissioned study: Museum of Modern Art
May 1986	Dr Danny O'Hare, DCU	Science, Technology, Children's Museum
Oct 1986	The Arts Council	Gallery of Modern Art
April 1987	Dr Charles Mollan, RDS	Children's Museum
June 1987	National Museum Ireland	National Folklife Museum
Oct 1987	Dept of An Taoiseach	Gallery of Modern Art, Folk and Science Museum
Nov 1987	Dept of An Taoiseach	Museum of Decorative Arts & Folklife
Jan 1988	RDS	Science Centre
May 1988	Department of Marine	Maritime Museum
June 1988	Irish Museums Trust	Creative Participatory Centre for Young People
Feb 1989	An Post	Irish Postal Museum
May 1989	National Museum Ireland	Museum of Postal History
Nov 1989	National Committee for Science	History of Science & Technology Museum
Jan 1990	Royal Irish Academy	History of Science & Technology Museum
Feb 1992	Stan Nielsen, EOLAS	Science Centre
Oct 1997	Irish Museum Modern Art	Second site for the IMMA
Mar 1998	Dorothy Walker	Proposals for a Scully Museum (Sean Scully)
Mar 1998	Maritime Institute of Ireland	Maritime Museum
Nov 1998	Report by CHDDA	Transport/Social History/Children's Museum
Apr 1999	Scroope Design	Proposal for Explorarium
Sept 1999	Dept of Environment	Social History oriented Transport Museum
Sept 2000	Chester Beatty Library	Museum of City of Dublin
1993–2000	DISCovery	Science Museum
Dec 2000	National Museum Ireland	Museum of Social History and Transport
Dec 2001	Dept Arts, Sport and Tourism	Museum of Dublin: A City and Port

(Department of Arts, Sport and Tourism.)

and marketing material produced by the CHDDA as a preferred facility for the museum complex. This is hardly surprising given that a large number of waterfront regeneration projects around the world have favoured the construction of similar facilities as a leisure-time facility. In Cardiff Bay, the Techniquest Science Discovery Centre has a central location on the waterfront and, more recently, Glasgow Science Centre has opened as a focus of the Clydeside Regeneration Project.

In Dublin, the perceived imminent development of a Science Centre as part of the Stack A regeneration led to the establishment of a lobbying team, the Dublin Interactive Science Centre (DISCovery) group, dedicated to setting up an interactive science centre in the city. They argued that without such a facility Dublin was falling behind many European capitals and, given the experience of other cities, including Cardiff, where the same architectural firm designed the docklands regeneration strategy, a science centre would be a very appropriate public attraction. The group first met in 1986 with the chairman of the CHDDA, Frank Benson, and was invited to submit a proposal for the development of a science centre in Stack A in June 1987. In a subsequent article published by Benson a stamp of approval appears to have been given to the project as he specifically outlines the content of the cultural facilities to include 'a folk museum, museum of science and museum of Art' (Benson, 1988). Following the surprise resignation of Benson from this post, further meetings were held with the CHDDA culminating in a second proposal submitted to the CHDDA Secretary, Gus MacAmhlaigh, in August 1991 that led to an invitation in 1993 to prepare a feasibility study. The report was produced in collaboration with Deloitte & Touche, and evaluated by EOLAS, an advisory and information centre based at Dublin City University (DCU), who considered it a worthy project but worried about the financial sustainability of the proposal. Their evaluation concluded that:

> Stack A is an excellent location for an ISC but the conversion and overhead costs associated with Stack A are high. It finds the DISC proposal attractive in concept but has doubts about the financial viability. Such doubts centre largely on the feasibility of obtaining the necessary Structural Funds and of raising the required sponsorship. Further attention to certain costings and to the estimated income from gate receipts would also be necessary. These doubts are very much intensified by the imminent establishment of Scientazia which seems likely to be a strong competitor for both markets and sponsorship.

Scientazia was another proposed interactive science centre for Dublin that argued for development on the grounds that in Europe only Ireland, Greece and Portugal did not have an interactive science centre, and in Britain alone 32 similar centres existed. Following their decision to withdraw from developing a full interactive science centre in the city, the DISCovery group obtained 'approval in principle' from the Custom House Docks Development Authority, subject to approval by the Board and agreement by the development company. The anticipated response was not forthcoming, and on 3 November 1993, Mahon Murphy of Deloitte & Touche, representing DISCovery, wrote to the development consortium stating that:

> Following many recent meetings with the Custom House Dock Authority and Brian Owens of Hardwicke, I was led to understand that Hardwicke would last month issue a letter to the CHDDA which would give the organizers of the Science Museum three months to raise the requisite funds for the project. Unfortunately, this letter was not issued ... I understand that there is now to be a further delay, while some alternative uses for the site are explored.

A letter in response from the developers suggested that they had not agreed to grant the group a three month option to raise funding and 'for legal and other reasons ... were not in a position to enter into discussions with any party on an option basis for the above building ... and therefore see little purpose in having such a meeting as your group are quite clearly seeking to have an option on the building for a given period'. The disagreement continued but, before it was satisfactorily resolved, the CHDDA published a new planning scheme for the extended Custom House Docks Area, favouring a complete re-zoning of Stack A. There is enough coincidence of timing to assume that this may have been one of the 'other reasons' why the development consortium would not allow any group, including the proponents of the science centre, an option on the building. At the time, the Hardwicke/British Land consortium had exclusive rights to the building and favoured the development of a shopping centre as the most profitable commercial option. It is highly probable that they were the key proponents behind the attempt to alter the zoning from cultural/amenity to 'all retail' in the 1994 Planning Scheme.

This change was questioned during a meeting between Gus MacAmhlaigh and Terry Durney of the CHDDA and DISCovery representatives in October 1994, who objected to the proposed designation of Stack A for retail only. In that meeting it emerged that, even though rezoning was being debated, the CHDDA had already granted Dublin Tourism a portion of Stack A to develop an attraction called 'Gulliver', based on Jonathan Swift's classic. In a letter to the authority of 5 November 1994, taking on board Dublin Tourism's

119 Artist's impression of Stack A redevelopment. (DDDA.)

120 Renovation commences 2000. (N. Moore.)

121 Conservation work on Stack A. (N. Moore.)

proposed attraction and the value of creating a mass of cultural attractions, Lewis Clohessy on behalf of the DISCovery team said:

> The DISCovery Group believe that our project is completely compatible with the spirit of the authority's original Planning Scheme of 4 June 1987 and with the conservation strategy for Stack A, that it would blend admirably with other cultural uses for this historic building and that, if a significant portion of Stack A which might otherwise have been used to educate, inform and entertain the public is 'lost' to retail uses, the integrity of the authority's own vision for the renewal of this unique quarter of central Dublin will have been seriously damaged.

They went on to say that DISCovery would be happy to share the space and management, marketing and other facilities with Dublin Tourism if the CHDDA would support their proposal. The science centre was not given approval to proceed, but the group was successful in fighting to retain the original museum zoning, and the CHDDA conceded in a follow-up meeting on 1 December 1994 that the authority 'has a legal obligation to provide a genuine Museum in the complex'. The 1994 Plan, finally approved in January 1995, stated that:

> Stack A is being given a special designation in that the authority will consider any use which is likely to achieve its conservation objective and encourage a lively use of the building, with preference for a cultural or major visitor attraction use.

The attempt to progress the procurement of a museum for Stack A beyond an aspirational stage was again blocked in November 1995, when the chairperson of the CHDDA, Professor Dervilla Donnelly, wrote to all interested parties informing them that proposals for Stack A would have to be resubmitted, pending a series of new guidelines and terms of reference. This indecision resulted in the loss of the Gulliver's Museum to Stack A, as Dublin Tourism withdrew their application. By early 1996 almost all the other components of the original Custom House Docks development were either complete or nearing completion, leaving Stack A as the remaining challenge, a derelict building in the middle of a new financial centre. Although Laura Magahy of Temple Bar Properties, seeing an opportunity for the rival city-centre urban renewal area, offered to house a new science centre and children's museum in the Poddle Building within Temple Bar, the DISCovery group declined the offer. For their purposes, Stack A was the preferred option, given that the development levy on commercial interests in the CHDDA as a condition of tenancy would amount to almost €1 million and was critical to the viability of the scheme.

This may have proved to be an unwise decision, given that a Government directive in 1996 forced the CHDDA not only to reject further representations by lobby groups but to freeze all plans for Stack A, pending a review of a new suggestion for the building by Dermot Desmond. He proposed constructing an ecosphere, a €216 million project comprising a multi-storey 80-metre high glazed pyramid over George's Dock containing an aquarium and a simulated forest through which people could wander and view the plants and animals of the world. The pyramid would be entered through Stack A, which was to be filled with shops and restaurants. Although ridiculed in some quarters as the 'Gorillas in the Mist' project, Desmond had already demonstrated his commitment to the project by spending over €1 million researching its feasibility and arguing that this could become a major millennium project for the city. In the minutes of a meeting with the relevant Government parties in February 1999, Desmond argued that 'he wanted it [the IFSC] to be a living centre which would not be dead in the evenings and at weekends, and that he has made this commitment to the IFSC'. An independent report by Peter Bacon commissioned by Bord Fáilte to assess the project gave it a 'cautious welcome' and argued that it would probably break even financially after five years. Its attractiveness to Bord Fáilte lay in the fact that it was one of the few major tourist attractions that could potentially assist in enticing families to holiday in Dublin. While Dermot Desmond had agreed personally to underwrite and contribute £25–£38 million towards the capital costs of the scheme, the public sector commitment to running the project was considered to be excessive. In early 2000, the Taoiseach, Bertie Ahern, wrote to Minister for Finance, Charlie McCreevy, stating that the new Dublin Docklands Development Authority were getting agitated by the delay in making a decision on Stack A. He urged the Minister to turn down the project, given that it had been under government consideration for four years and Stack A formed an integral part of the ecosphere proposal. The DDDA, which had taken over from the CHDDA in 1997, could not move on its conservation and regeneration objectives until a final decision was made on the future use of the building.

There was no certainty over the building until the Government cleared the way with a decision on Desmond's proposal, yet other groups including DISCovery continued to lobby the new authority in the hope that, if the Desmond proposal was rejected, their projects would be considered a feasible alternative. In a letter of September 1998 Peter Coyne, Chief Executive of the DDDA, indicated quite clearly to DISCovery that it was unlikely the science centre would prove a good fit for Stack A and that instead it should consider locating elsewhere within docklands:

> The key thing that Stack A requires is certainty of a fairly immediate delivery of a quality museum project. The National Irish Science Centre project is exciting but, I am sure you will concede, is presently uncertain in its funding. I do think that it is an excellent idea for the Irish Science Centre and the Pigeon House project [another science centre proposal] to get together, and I hope that a project emerges, perhaps taking advantage of the striking setting of the Pigeon House.

For the first time it appeared that the DDDA were beginning to generate an alternative vision for the warehouse that did not conform to the original plans. However, it was not the only institution to argue the unsuitability of a science centre in this building. In an internal memorandum of the Department of

Arts, Heritage, Gaeltacht and the Islands, a Principal Officer suggests to the Minister that 'the department would strongly support the need for an ISC, but Stack A with its listed status would not in our view be the most suitable building or large enough for a proper science museum'.

To progress development urgently on the site, the docklands authority made a final call for proposals in March 2001. This was followed up by correspondence from Gráinne Hollywood, its Director of Property, indicating the amount of floor space that would be available to a museum project. Dramatically reduced from the initial plans for a 12,170 square metre museum and amenity centre in 1987, Hollywood indicated that the amount of space now available for this element of the project would total only 2,654 square metres on the ground floor and 1,834 square metres in the basement. It is hard to see how the authority arrived at this decision given that a Government commitment was made as late as 20 January 1998, following the establishment of the DDDA, that 'the museum element of Stack A will amount to 5,760 square metres. A national folk museum will be provided at Stack B, on an area of 2,230 square metres, adjacent to Stack A'. Not only had the original floor space in Stack A been greatly reduced between 1998 and 2001, but Stack B had also been demolished. The difficulties that the DDDA decision generated for proponents of various museum facilities is made clear in another internal briefing note from the Department of Arts, Heritage, Gaeltacht and the Islands of July 2001 remarking that

> the space on offer from DDDA in Stack A has ranged in their correspondence from a low of 36,900 square feet [3,428 square metres] for the cultural facility to a high of 48,000 square feet [4,459 square metres]… If the final concept of the museum put forward by us does not appeal to their commercial sense, they will oppose it.

The complete autonomy of the DDDA facilitated these changes, and its decision to terminate the Master Project Agreement with the Hardwicke/British Land consortium in September 2001 ensured that Stack A would be redeveloped in the manner considered most appropriate by the DDDA alone. Having considered the final submissions, Peter Coyne wrote to museum proponents in February 2002, notifying them that the preference of the DDDA was to work with the Department of Arts, Heritage, Gaeltacht and the Islands on a proposed Museum of Dublin. In a letter to the DISCovery group (and running contrary to the outstanding successes of similar amenities at the @Bristol Science Centre, Techniquest in Cardiff and the W5 Centre at the Odyssey on Belfast's Laganside), the decision of the DDDA was justified by arguing that the visitor projections of 300,000 per year were too high and that the science museum proposal was not financially sound.

What has since emerged is that the following month, on 27 March 2002, Peter Coyne also wrote to Paul Haran (now Principal of UCD College of Business and Law) in the Department of Enterprise, Trade and Employment, suggesting to him that locating a purpose-built science centre on the other side of George's Dock might be of interest, a site that is currently earmarked for the new Abbey Theatre. This would suggest that the DDDA were not opposed to the idea of an interactive science centre *per se*, but rather did not want to see it occupy Stack A. This idea was further developed through a report commissioned from Murray O'Laoire architects by the DDDA in summer 2002, for a museum of about 650 square metres. It concluded that 'the project should have a solid future in terms of attracting visitors and combined with the proposals for Stack A is easy to envisage a 'critical mass' of attractions which would warrant long and return visits'. Although this was sent to the Office of Science and Technology (OST) by Peter Coyne, who indicated that he would be pleased to work with the office to further the proposals, nothing more emerged. Simultaneously, the OST were in contact with DISCovery in an attempt to raise funds for their proposal. In a letter of 26 July 2002, Mr Martin Shanagher stated that:

> Government approval in principle for the funding of a science centre will be required as the next step to progress the matter. The OST is working to obtain this approval, however it is unlikely that there will be any decision before the autumn.

What this did not make clear was whether the funding was being sought for an interactive science centre in general or for the DISCovery project specifically. What is more unusual is that while the OST knew of the Murray O'Laoire report and was concurrently in talks with DISCovery, it never informed them of the alternate proposal. In a letter to them, the Office of Science and Technology argue that they 'did not enter into discussions with the DISCovery Group in relation to the other site in the IFSC as the DISCovery Group proposal related specifically to Stack A'. Yet Rose Kevany, the chairperson of that group, believes that:

Not being appraised of this development meant that we were deprived of the opportunity to raise funds for a science museum in the IFSC, an opportunity which was subsequently afforded to the Children's Museum group by providing a building at Heuston Gate for which they were authorized to raise funds of approximately €18 to €20 million.

The OST counter-argue that even if they had provided funding 'the decision not to locate a science centre in the Dublin Docklands Area was entirely a matter for the Dublin Docklands Development Authority and not one over which the OST had any influence'. While the final chapter had closed on the sixteen-year attempt to develop a Science Centre in this flagship of nineteenth century engineering, a much shorter one was just about to open.

The Museum of Dublin

When the docklands authority contacted other competitors in February 2002 to inform them of their decision to work with the Department of Arts, Sport and Tourism on the development of a Museum of Dublin History, it seemed almost certain that the final piece of the jigsaw in the Custom House Docks project had been found. The proposal originated in 1999 when the Minister of the Environment, Noel Dempsey, had suggested the development of a social history and transport museum. This proposal was further refined by the Department of Arts, Sport and Tourism and a decision was made to focus the project on a much narrower theme, *Dublin: Its History as a City and Port*. One of the reasons that this idea appealed to the Minister was the fact that Dublin City Council had already established a Working Commission for the city of Dublin Museum and had progressed discussions with the docklands authority. The initial submission requested that the allocated floor area for the museum would need to be increased to 6,600 sq metres to guarantee the viability of the project, while the amenity should also have operational independence and total autonomy from the DDDA, and pay a nominal rent. The rationale given by one of the members of this Commission, Professor Loughlin Kealy of UCD, was that he had 'reservations about the concept of the museum being pushed into a space left over from something else, i.e. the various retail outlets and commercial ventures envisaged by the DDDA'. One of the other members argued that he believed the DDDA had an attitude problem and failed to understand that museums need to be supported and are not profit-making institutions, while Philip Maguire, Assistant City Manager at the time, explained that although the DDDA 'had a preference for a cultural usage … commercial operation was at the core of their activities'. A letter of 11 November 2001 issued by the DDDA in response to these queries reinforced this viewpoint, stating that there was no way these requests could be granted and 'that there is no possibility of it extending to 5,574 square metres or to taking over the total ground floor area of Stack A'. In correspondence earlier that month with the Department of Arts, Heritage, Gaeltacht and the Islands, the DDDA had also reminded Chris Flynn, a Principal Officer in that Department, of the importance of the development levy and that 'if there is no museum in Stack A, this very valuable rental stream which was negotiated by the authority and its predecessors will not be collectable and will be lost'. The immediate impact was a government decision in December 2001 to authorize a feasibility study and gauge how the project might be progressed. Meanwhile, a sense of urgency seemed to be growing as Peter Coyne warned the Department on 21 January 2002 that, unless the feasibility study was undertaken rapidly, 'the opportunity to consider Stack A as the home for the Museum of Dublin may be lost'. At the same time as the Government was being forced to fast-track the study and the DDDA was warning that the future of the project could hang in the balance, the latter were also writing to other museum candidates turning down their proposals and implying that the Museum of Dublin was almost a certainty. The power of the DDDA and the pressure that they appear to have been able to bring to bear on the Government is extraordinary. In statutory terms the authority is supposed to be answerable to and controlled by the Minister of the Environment and by extension the Government, rather than the contrary.

A call for tenders to undertake the feasibility study was placed in the national press on 11 February 2002 and, from the four short-listed candidates, Event Ireland was awarded the contract. Although they were due to report to the Minister by 30 April 2002 with their findings, events took a very unusual turn when the DDDA issued its own advertisement in early April for tenders to assess the feasibility of a Museum of Dublin. The similarities with the terms of reference that the Department had issued two months earlier were uncanny, and the situation was made even stranger by the fact that the DDDA had known exactly what the Department had been doing to move the project along:

> When the Department, Dublin City Council and the DDDA met to agree the terms of reference of the Department's consultancy, no

indication was given by the DDDA that this was in the offing. Of particular interest and concern is the 'Museum Project success criteria' set down in the DDDA document on page three. In my view if the DDDA require a footfall of at least between 350,000 and 500,000 visitors on an annual basis as being one of the success criteria to be achieved quickly, then from their perspective the project is not feasible.
(Internal memorandum, Department of Arts, Sport and Tourism, 8 April 2002.)

Given that the National Museums in Dublin receive annual visitor numbers well below this estimate (the most popular attraction at Kildare Street receives approximately 275,000 visitors per annum) the commitment of the DDDA to any museum facility began to be called into question. In the tender document produced by the Authority, one of the consultancy objectives states that 'the authority does not want the consultant to seek to prove that the project success criteria can be met – it wants a robust and honest appraisal of whether that is readily achievable'. This could be read as the authority requiring the consultants to reach a negative appraisal right from the start. The Event Ireland report for the Government concluded that the project was 'likely to be feasible, practical and sustainable as well as making a valuable contribution to Dublin's self-esteem, cultural life and tourism product'. They went on to say that the breakeven figure for the Museum would be 120,000 visitors annually at a rate of €6.50 per person, but that this could be increased to 200,000 visitors with little difficulty, given well-developed marketing. Yet the consultants appointed by the authority, Locum Destination/At Large, concluded contradictorily that visitor numbers were more likely to be around 100,000 visitors per year and that they would recommend a contemporary art facility for the building. This may have been why the City Arts Centre were also approached by the DDDA during this time, but they could not raise the revenue to provide an arts facility in Stack A. Summarizing the DDDA report for the Minister for Arts, Sport and Tourism, a senior official strongly rejected the approach of the DDDA and questioned their motives, stating that:

It is the view of this division that setting this target range for a brand new museum is tantamount to vetoing the concept. The reasoning and the basis for the conclusion in the Locum Report that a Museum of Dublin would not be viable are not persuasive in our opinion.

Unsurprisingly, following this prolonged exchange, the DDDA Board concluded negotiations with the Department in August 2002. On 26 September 2002, Peter Coyne wrote on behalf of the DDDA to all relevant museum parties stating that:

> In appraising options for the 'cultural' element a broad range of proposals was considered, and in particular, several museum type proposals were put to the authority. This kind of operation had been anticipated by the Custom House Docks Development Authority who had arranged for a subvention payment for an IFSC based museum, in IFSC leases. The contribution would equate to circa 1 million euro per annum if charged. In the final analysis, the various museum type proposals were felt to be insufficient in their power to attract visitors when operating in the Stack A centre, or were otherwise a poor fit to the overall project, or to the open bright form of the building, and none was selected. As it was, no museum proposal was found to be secure financially even with the benefit of the available subvention.

Even though the actions of the DDDA are indeed questionable in many respects, it is also the case that the Government may have found it difficult to source the requisite funds if the Board had decided to proceed with the Museum. In December 2002, the Government made a final decision to 'withdraw from further involvement with Stack A and leave the DDDA free to pursue the type of cultural facility that meets their requirement'. The final blow to obtaining a wide range of uses on the Custom House Docks site was dealt. The Dublin docklands project is now one of the only international waterfront regeneration projects without a major museum or similar attraction at its heart. In Rotterdam, a maritime museum is a central part of the redevelopment programme while, as mentioned earlier, the South Street Seaport Museum Trust in New York has been a central part of renewal plans.

Progress was eventually made, and the restoration of Stack A is now complete, following a €50 million building conservation project. Completed in August 2004, the type of facility finally developed is a small events and high-end retailing venue. Now known and marketed as chq, Stack A was designed to become a luxury shopping quarter, with five restaurants and 1,404 square metres of event and exhibition space. A new glass frontage on to the Liffey and glass wings facing on to George's Dock have been added to the façade of Stack A (Figure 123). The result has been a physical reopening of this

122 *chq* brand applied to Stack A, 2005. (J. Brady.)

123 Renovated Stack A, 2005. (J. Brady.)

124 Stack A and George's Dock, 2005. (J. Brady.)

part of the Custom House Docks site to the river and the city. Whether these facilities will achieve the original objectives of drawing people into the area after hours is unclear, and letting of the premises has had a number of false starts. The homeware retailer, Meadows & Byrne, agreed terms in July 2006 to become the anchor tenant for *chq*, following the decision of Harvey Nichols to locate in Dundrum rather than docklands. More recently Ely Restaurant and Wine Bar have opened a successful and popular operation, and it is hoped that others will follow suit.

One concern is that because of the marketing and nature of the project, the facilities provided and the kind of clientele it is aiming to attract, this amenity will lose its potential to entertain and inform the general public about a much-forgotten and ignored part of the heritage of Dublin. Additionally, doubts have been raised as to the viability of this kind of development at this location, and interesting comparisons have been made with the fate of London's failed Tobacco Dock development in Wapping. Part of the original plans for a shopping and entertainment centre in London docklands, the Tobacco Dock warehouse went into decline very soon after opening. It has been largely unoccupied since its closure and is only sometimes used for promotional events. Since 2003, the building has been on a 'Buildings at Risk' register compiled by the English Heritage Trust and is subject to new proposals for a mixed-use development. Coincidentally the recent proposals for Stack A in Dublin docklands, which was designed by the same architect and built six years after Tobacco Dock, are very similar.

The difficulties and controversies surrounding the redevelopment of Stack A exemplify that the 'built heritage under various renewal schemes has been seen as a commodity valued in economic and tourism terms rather than for its cultural or social significance' (McManus, 2005, p. 242). The prolonged debate regarding the future of this building has highlighted the reluctance on the part of development authorities to engage in the conservation of a building and promote it for heritage reasons alone. In this case, only when an economically profitable vision of its future became dominant did the DDDA invest in conservation work. This is not unusual, as it would appear that it is often the broader cultural and economic context that drives our attitude towards heritage. Evidence would suggest that whether we value particular popular memories, such as those of the dockers, depends on whether or not they can be constructed as a marketable story. Very often the potential of a place to tell us something about the lives of ordinary individuals in the past is less valued than those places that represent the extraordinary. So, while we revel in the history and elegant lifestyles of the elite through visits to places like the National Museum of Decorative Arts at Collins Barracks, the old Parliament at the Bank of Ireland on College Green, town palaces like Powerscourt House and other monumental buildings throughout the city, where do we hear or see the stories of ordinary Dubliners who witnessed some of the most dramatic events in Irish history from their tenement homes?

In Dublin docklands, this disjuncture between the marketing machine and the memory of local residents is even more apparent. Community groups argue that their heritage, memory and identity is being devalued in the construction of a new place identity. That identity is based on the prominent position of the Dublin docklands in international finance networks and fails to recognize the less-polished 'everyday' history of docklands, and diminishes the significance of the relics of the maritime heyday. The North Wall Community Association has argued that:

> There is nothing to reflect the culture or history of the area. Sadly Stack A, which was ring-fenced back in 1992 for that specific purpose is to become an area of bars and restaurants. When you look at the Albert Docks in Liverpool with its Maritime Museum, reflecting the great

seafaring history and tradition of Liverpool [...] you can get a sense of what could have been achieved in Stack A. Instead very shortly the developer driven concrete jungle will be complete and the sands of time will wash away the memory of the history of the Dockland [...] who will recall the Guinness boats tied up alongside Custom House Quay? [...] Who will recall the men who worked on the docks, the men who dug the coal boats with their no. 7 shovels? [...] The lives of the ordinary people and their daily struggle to survive [...] who will remember? Who will tell their story?
(North Wall Community Newsletter, Easter 2005.)

Instead, Stack A is being marketed in a highly selective manner and, rather than providing a place for people to congregate, *chq* may become a new symbol of segregation and exclusivity in an already divided district. The exclusive nature of this development, and its incongruence with the history of the area and the building itself, was foreseen in the feasibility report provided to the Government by Event Ireland in 2002. They declared that this combination of activities:

will not assist a policy of social inclusion if the retailing environment appears hostile to lower income groups, and contrasted it with the Museum of Dublin that aimed to have a socially-inclusive marketing policy and will not attempt to confine itself to an upmarket audience.
(Event Ireland, 2002, p. 35.)

Returning to the original debates in modern Ireland about the relative balance that should be afforded to heritage and development priorities, Senator Gemma Hussey during the Wood Quay parliamentary debates in December 1978 urged that:

Just because the whole business of Wood Quay has become a cause célèbre it should not be allowed to be made into any kind of a political football or made into a matter of somebody's pride or bad feeling or the subject of some kind of gamesmanship.
(Seanad Debates, 6 December 1978.)

Similar comments could be made regarding the redevelopment of Stack A. It would appear that a massive opportunity has been lost to embrace the maritime and industrial past of Dublin as a city and port. The lack of any firm government control on development at this site has resulted in a zero-sum game from both an economic and heritage perspective. Although a dichotomy did emerge in the last decade between those who wished to see the facility conserved for public access and cultural use and those who preferred a more commercial option, the bottom line is that the building is only now, after twenty years, beginning to generate revenue. The ability of a State-established development agency to browbeat a Government into withdrawing a proposal for a heritage facility in such an important location is astonishing, and demonstrates a reluctance on the part of the Legislature to take heritage seriously. The result has been a displacement of identity for local communities, as their memories and the goals of contemporary place marketing are diametrically opposed.

The politics of development

Although the Stack A story represents a long-running controversy stretching from the very beginnings of the docklands renewal project in the late 1980s, more recent phases of redevelopment have also resulted in bitter and acrimonious exchanges. While disagreements over the fate of Stack A represented philosophical differences over the focus of regeneration, disputes over how redevelopment would occur at Spencer Dock, east of the IFSC in the north docklands (Figure 125), had more critical repercussions. The plans proposed in 1999, the cause of so much heated-debate, have been perceived as an attempt by some groups to re-engineer not only the physical, but also the social geography of the northern docklands through marginalization and displacement. One local representative, Councillor Gerry Breen, in a statement to the Planning Appeals Board on 23 February 2000, summed the situation up as 'ironic … in Dean Swift's Dublin, Spencer Dock is proposing a model of Lilliput. But we are the Lilliputians.'

Part of the difficulty may be traced to the actual institutional framework put in place by central government. When it was established the DDDA, unlike its predecessor the CHDDA, was not awarded planning powers for the entire 526-hectare area under its remit. Instead, the authority has responsibility for planning within the area in conjunction with Dublin City Council, but planning is only fast-tracked or exempted from the traditional planning process in specially designated sites. This complex institutional landscape and the difficulty in co-ordinating a strategic vision for the district were partly responsible for the delay in instigating redevelopment at Spencer Dock. But perhaps the greatest delays in redevelopment came from the fact that the rejuvenation of this site, one of the largest and most strategic sites within docklands, had also become associated with a more complex political controversy regarding the development of a National Conference Centre (NCC) as a major flagship project.

In many countries, large-scale urban renewal projects have used flagship projects as both a way of attracting investment but also as the core element in re-branding or marketing exercises. In places like Canary Wharf in London or the Potsdamerplatz in Berlin, government intervention in the form of tax incentives, public-private partnership or a stream-lined planning system has facilitated the emergence of these iconic projects. An unforeseen or perhaps even a calculated result of such action has been the politicization of particular projects. Development schemes have become closely linked with particular political personalities, and acquired disproportionate symbolic significance. One of the most renowned examples was the Millennium Dome in London where the delivery of the project and the standing of prominent political personalities became very closely intertwined. Many have argued that this connection was partly responsible for the subsequent failure of the scheme as it was considered in many circles to be an 'entirely political project' (*The Guardian*, 16 June 2000). A similar criticism could perhaps have been levelled at the early development of the Custom House Docks in Dublin and in particular with the development of the IFSC, as previously discussed.

In the context of a more recent project on a similar scale to the proposed National Conference Centre in Dublin, a similar politicization of urban regeneration became apparent in Cardiff in the early 1990s regarding the Opera House project, proposed as part of the Cardiff Bay regeneration programme. Following an extensive competitive tendering process, concluded in 1994, Zaha Hadid – an Iraqi-national based in London – was chosen as the winning architect for the scheme. Her design garnered equal amounts of support from some quarters, and concern from others, who argued that the design was inappropriate for Cardiff Bay. The Opera House was designed to become the home of the Welsh National Opera, attracting criticism, particularly from the tabloid media, of elitism. Although proposed as one of the twelve great Millennium projects to be funded from National Lottery money, neither the local authority nor the Cardiff Bay Development Corporation fully supported the proposal because of the controversy surrounding the project. Even the Welsh Office, who had committed £2

125 Spencer Dock environs, 1994. (DDDA.)

126 Spencer Dock environs, 2005. (J. Brady.)

127 Spencer Dock environs, 2005. (J. Brady.)

million sterling to progress the project, procrastinated when its support was most critical (Crickhowell, 1997). When the Millennium Commission denied funding for the project, its death-knell sounded. Lord Crickhowell (former Secretary of State for Wales) has argued the merits of such a scheme and, in deciphering why the decision was taken not to fund the Opera House, he draws clear links between the power of the media, politics and the development project:

> The Commission's interest became concentrated on worthy environmental schemes which did not invite press charges of elitism ... some people believe that ... the Commission caved in as a result of a populist clamour against elitism and the hostility of the Sun newspaper. It is suggested that a political decision was taken that, after previous rows, the easy way out would be to sacrifice the Welsh.
> (Crickhowell, 1997, p. 151.)

The Opera House Trust set up to oversee the delivery of this project was disbanded in 1996 amid local authority promises to develop a smaller, commercially-led, more populist theatre. This has taken the form of the Wales Millennium Centre, a multi-purpose entertainment centre suitable for musicals and operas, constructed at a cost of £86 million, and which opened in

2003. It is the new home of the Welsh National Opera but also the base for arts companies, such as the National Dance Company of Wales. Contrary to press reports, the facilities included in this proposal are almost identical to the Hadid-Opera House project, yet this proposal received Millennium Commission funding. Prior to a decision on funding being made, Lord Crickhowell questioned how the Millennium Centre could even be considered given that the business plan could in no way be considered as solid as the Opera House proposal. He concluded that 'if they do back the project it will be clear that the criticism of the Cardiff Bay Opera House Trust's business plan was no more than a smoke-screen, and that the decision to reject our bid was largely political' (Crickhowell, 1997, p. 167).

This clear politicization of the urban development process in Cardiff has resonance with the fiasco that has characterized the attempted procurement of a National Conference Centre in Dublin. As with all major public projects, central government launched a tendering process for this scheme in 1989. It was unsuccessful, due to the weak economic and property market at the time, and a second bid launched in 1995 was abandoned by the Minister for Sport and Tourism. This followed widespread reporting that the Royal Dublin Society (RDS) in Ballsbridge was the preferred site for the project. As the RDS is a quasi-public institution, the organization would have succeeded in drawing down 75 per cent funding from the European Union for the project, in contrast with the maximum 25 per cent funding that could be allocated to a private developer. The anti-competitive nature of this resulted in a formal complaint to the European Commission by the Carlton Group, a private consortium with a preferred site on O'Connell Street, who was also among the bidders to construct the Conference Centre.

Amid much public debate, a third competition was launched by Bord Fáilte in 1998, and five schemes were short-listed for consideration. Treasury Holdings proposed building the Conference Centre at the CIÉ marshalling yards at Spencer Dock; the RDS argued for a site in Ballsbridge; the Office of Public Works favoured a location on Infirmary Road; the Sonas consortium had intentions to build on the old Phoenix Park racecourse; and the Anna Livia consortium proposed developing at the Grand Canal Docks (Figure 128). Following consideration of all proposals, Bord Fáilte recommended the acceptance of a public-private joint venture project at Spencer Dock in the north docklands, which the Minister for Tourism, Sport and Recreation, Dr McDaid, was happy to accept, given the political importance of the proposed project

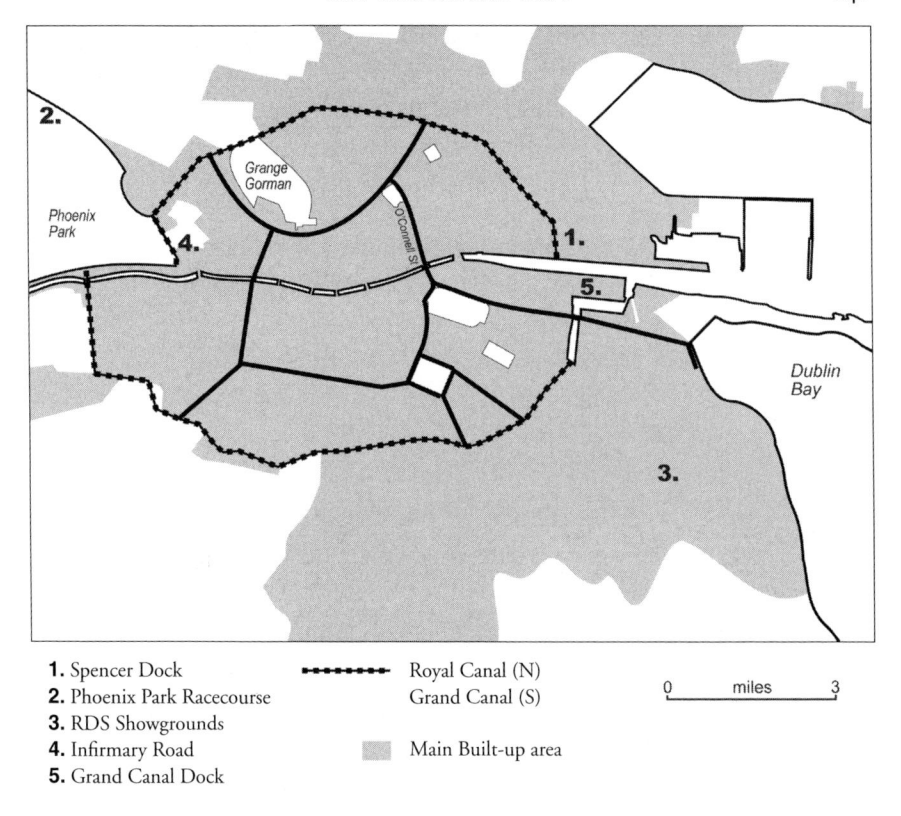

1. Spencer Dock
2. Phoenix Park Racecourse
3. RDS Showgrounds
4. Infirmary Road
5. Grand Canal Dock

Royal Canal (N)
Grand Canal (S)

Main Built-up area

128 Proposed locations for the National Conference Centre. (N. Moore.)

> The Government views the proposed development as a very important flagship project and a major addition to the tourism industry's physical infrastructure. In particular, the project has the potential to generate significant levels of foreign tourist revenue, much of it during the off-peak season.
> (Government press release, 16 June 1998.)

Unusually, this joint venture, or public-private partnership to use the parlance that is currently favoured among policy analysts, was being led by CIÉ, the national transport company, and the private developer, Treasury Holdings. Questions emerged and were never satisfactorily answered publicly as to how a public-sector agency and a private-sector developer had reached a development deal without the knowledge of the responsible Minister or the Government. As late as July 2000, the Minister for Transport admitted that officials in her department had no idea how, or what kind of, agreement was reached. The local representative, Tony Gregory, believes that the deal was never really exposed to the extent it should have been. How the State allowed that to happen, or how the Ministers allowed that to happen under their noses, I have no idea. I only discovered afterwards that they couldn't get out of it, but whether they wanted to or not is another day's work. They claim they wanted to get out of it.

This lack of both transparency and accountability was heavily criticized by, among others, Patricia McKenna (Green Party, MEP) who stated that:

> CIÉ may own the land, but don't forget CIÉ is a public body. At the end of the day, the public really own the land. And why should a huge profit be made without any proper transparency?
> (*Questions and Answers*, RTÉ television, 17 July 2000.)

Even before the exact content of the proposed development became known, controversy was emerging over how, and under what terms, the development deal had been struck. In the context of what later transpired and the escalation of conflict between a range of groups over this site, it seems clear that profit maximization was the key motivating factor for both CIÉ and Treasury Holdings. Interestingly, prior to the submission by the consortium of a development proposal for planning permission, the Dublin Docklands Development Authority had already shown interest in acquiring the rail yards from CIÉ, but at a significantly lower price than market value. The State transport company rejected this deal, favouring the private partnership arrangement with Treasury Holdings which would ensure that CIÉ retained freehold of the site. Had the site been sold to the DDDA, or any other group, proceeds of the sale would have gone straight to the national Exchequer. By entering a leasing arrangement, CIÉ would accrue annual rental income that in the long-term would amount to a much larger sum than a one-off sale (McDonald, 2000). Whether or not this is appropriate behaviour for a public sector organization without the apparent knowledge of the relevant Minister is questionable but the rationale behind the decision and its subsequent implications for urban planning highlight the fraught relationship between central government and many state agencies in Ireland. How agencies are financed is a crucial determinant in the process and outcome of decision-making, particularly in times of economic uncertainty. If government funding for development is uncertain, alternative ways must be found to raise future capital investment and that may well have been the driving force behind the emergence of the initial super-project at Spencer Dock.

Spencer Dock: the proposed development

Spencer Dock is a 20.85-hectare site located on the CIÉ marshalling yards at North Wall Quay and is twice the size of Temple Bar. The scheme proposed by Treasury Holdings for Spencer Dock, the largest ever submitted to an Irish planning authority, comprised 557,418 square metres, more than the total at Canary Wharf in London (520,257 square metres). The scale of the proposal is illustrated by the fact that many buildings would be built at a height of 95 metres, some 36 metres higher than nearby Liberty Hall, one of the tallest buildings in Dublin.

The proposal of the Spencer Dock Development Company (made up of CIÉ and Treasury Holdings) comprised the 2,000-seat National Conference Centre that had provided the initial impetus for development; 3,012 apartments in eleven blocks; nine office blocks; two hotels; and a technology park designed and operated in association with Trinity College Dublin for knowledge-based enterprises. A total of twenty-seven buildings would be constructed, many over twenty storeys high (Figure 130). The project would significantly alter the socio-economic and demographic profile of this part of the city through the creation of 11,000 permanent jobs and the attraction of 9,000 new residents. In terms of the heritage and aesthetic potential of the area, the planning scheme proposed to retain all listed buildings and, as is a common feature of urban design in similar schemes globally, the height of the buildings was to be stepped down towards the river. While these all seem like positive and creative ideas and would perhaps be perfect for a greenfield site on the edge of the city, the urban context and wider implications of the scheme were not considered. The proposal would appear to have been developed on a figurative blank canvas rather than considering the site in the context of its relationship to the city as a whole and to the surrounding neighbourhoods.

Although the consortium proposed to begin works in September 1998, a number of developments at local, national and EU-level conspired against this. Uncertainty existed over whether the National Conference Centre element of the project would be entitled to full urban-renewal incentives, given the contemporaneous debate over State aid within the European Commission. In addition, relations between the developers and Dublin Docklands Development Authority became fraught. In order to fast-track development by sanctioning a Section 25 Planning Scheme for the area, which would effectively have exempted the project from the usual planning channels,

129 Advertising banner for Spencer Dock development. (J. Brady.)

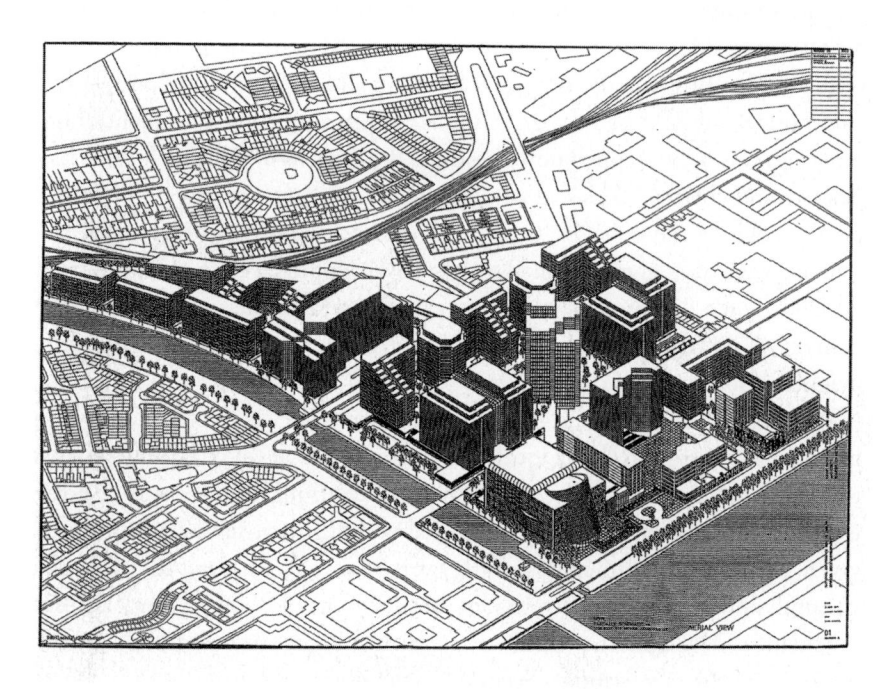

130 Schematic drawing of the original Spencer Dock proposal.
(Planning Application Ref: 0600/99.)

THE CONTESTED CITY

131 The Spencer Dock development in its context. (Environmental Impact Statement, Planning Ref: 0600/99.)

132 Older residential development, Upper Mayor Street and environs. (J. Brady.)

the DDDA requested further information on the detailed content of the proposal to ensure compliance with the overall docklands Master Plan. In a draft master plan that they had devised for Spencer Dock, the docklands authority had capped development on the site at 325,160 square metres and stipulated that no building, with the exception of the landmark National Conference Centre, should exceed 39.6 metres in height, shattering the plans of the developers, who were much more ambitious in scope and scale.

The consortium refused to scale down the project to the extent proposed by the DDDA, and it abandoned attempts to obtain a Section 25 certificate, opting instead to apply for planning permission to Dublin City Council through the regular planning channels. Full planning permission for the first phase of the scheme, comprising the conference centre, one 17-storey block of 223 apartments, four retail units and underground parking for 962 cars, and outline planning permission for the remainder was applied for. In October 1998, outline planning permission was granted pending the submission of additional information, as the authorities did not consider the quantity and density of development proposed to subsidize the National Conference Centre as desirable. Because of the fragility of the National Conference Centre within this overall debate, Bord Fáilte attempted to intervene and clarify issues in order to progress the project. In a letter dated 14 October 1998 to the Spencer Dock Development Company, perhaps designed to generate some sort of compromise, Bord Fáilte stated:

> The total development area is 28 acres of which 3.3 acres is dedicated to a stand alone National Conference Centre. The development area within the remaining 25 acres, proposed to support the National Conference Centre is 4 million square feet. The contribution of this development towards the National Conference Centre is represented as being:
> £3.30 per square foot on commercial development per annum
> £1.64 per square foot on residential development per annum
> £5 average per night per occupied room from hotel developments
> Spencer Dock Development Company is liable under guarantee for any funding shortfalls in excess as banker of last resort.

The consortium refused to assume this risk and engaged in one of the greatest ever speculative gambles in Irish planning history. Even in the face of objections from the Dublin Docklands Development Authority regarding the location of the National Conference Centre building within the overall site, the developers continued their drive to secure 557,418 square metres of development. In what might be interpreted as an attempt on the part of the developer to hold the DDDA and in effect the national government to ransom, due to the high profile nature of the National Conference Centre, work was suspended on-site until the DDDA objection was withdrawn. In November 1998, the DDDA guaranteed the developers that they would not object to enabling works at the National Conference Centre site, but they continued to raise objections regarding the proposed location of the building, because it would inhibit the development of a linear park as proposed in the docklands Master Plan. With the support of the Minister for the Environment, the DDDA argued, in the knowledge that the development consortium could try to use this objection as a way of developing CIÉ land without constructing the National Conference Centre, for the building to be simply moved 15.2 metres further east. A letter to Treasury Holdings by Peter Coyne, quoted by McDonald (2000, p. 79), warned against attempting to renege on the commitment to the Conference Centre project, and stated that the DDDA were

> very concerned that you may not be serious in your endeavour to secure the conference centre as you effectively disregarded our advice to you as to how best to proceed with the process. Furthermore, you have stated on more than one occasion that you would be much better off if the conference centre project were to fail and you were to develop CIÉ's land without it. That may be a simple statement of fact, but the implication is worrying.

Dermot Dwyer, a representative of Treasury Holdings, argues that they were 'never obliged to develop the National Conference Centre on this site and had simply entered a contractual arrangement with CIÉ for a mixed-use development'. This is questionable, given their successful bid for the project at this location. In a last-ditch attempt to prevent the collapse of the National Conference Centre project yet again, Dublin Corporation established a project team headed by the City Planner, to devise a set of principles to guide development on the site. These laid down the parameters within which the development proposal would be granted planning permission. Meanwhile, relations between the consortium and the docklands authority became even more tense, with the DDDA arguing that when CIÉ entered the joint venture they allegedly agreed to the provision of two public parks and a third-level campus but neither of these featured in the scheme submitted for planning permission. To ensure that this latter facility did not get embroiled in the planning controversy and thereby get delayed, the DDDA gave a site within the IFSC II development to the National College of Ireland to fulfil the desire for a third-level campus within the north docklands, as described in the previous chapter. The DDDA also objected that the requirement of the docklands Master Plan for 20 per cent social and affordable housing was not being met by the Treasury/CIÉ proposal, and that basic urban design principles regarding overshadowing and overlooking were completely ignored.

Demonstrating the extent of confusion surrounding the entire project, objections lodged against the project even came from within the development partnership. At the time, CIÉ was the umbrella group for a number of different public transport companies, including Dublin Bus, Irish Rail and the LUAS (light-rail) system. The project team working on the LUAS objected to the Treasury Holdings proposals on the grounds that the development failed to provide adequate space within which to provide an efficient LUAS extension to the Point Depot in the future. However, on the instructions of their parent company the objection was rapidly withdrawn, but demands from city planners that CIÉ would have to provide very specific commitments regarding the public transport facilities to service the new centre further strained the development partnership (*Irish Times*, 14 May 1999).

Other key concerns included the huge volume of ancillary development within the overall site that would negate the landmark status of the National Conference Centre (NCC). Complicating matters even further was a ruling from Brussels that, in order to benefit from EU aid, the National Conference Centre would have to be completed and operational by January 2001, a deadline that the developers thought they could manipulate in their favour. The controversy surrounding the project continued to grow amid allegations against the developers of coercion and lack of consultation with local groups. The development company attempted to neutralize the risk of objections from local residents in Upper Mayor Street, a cul-de-sac of six terraced houses particularly affected by the development, before the submission of proposals. Marie O'Reilly, chairperson of the North Port Dwellers Association and long-term resident at Mayor Street, told the *Irish Times* (24 February 1999) of attempts by developers to bribe residents to either leave the area or accept the proposals. She stated that residents were offered three options to convince them to leave, one of which included a once-off lump sum plus the market value of their houses. The tokenistic community consultation that had taken place only eight days prior to the submission of development plans to the local authority alarmed residents and led them to question what the future might hold. Explaining their reservations regarding the development, Joe Mooney, an East Wall resident, reasonably argued that

> we're opposed to the Spencer Dock plan not because we're against progress, but because of the negative impact it would have on our community ... This could destroy us and we haven't been consulted.
> (*Sunday Business Post*, 27 February 2000.)

Rejecting the arguments of developers that locals were resisting change at any cost, Marie O'Reilly (Mayor Street resident) maintained that:

> While we welcome planning and development through consultation and partnership, we do not need Spencer Dock to bring in a 'thriving and vibrant community'. The existing community is already that and has been since 1847.
> (Letter to the Editor, *Irish Times*, 28 January 2000.)

She also contextualized the problem noting that:

> My husband's family have lived in this house since 1847 and worked for the railways. We feel cheated ... It is ironic to see CIÉ now part of a consortium attempting to get rid of former railway worker's housing.
> (*Sunday Business Post*, 20 February 2000.)

Planning in a democratic state?

Because of the numerous institutions involved, the range of arguments made on both side of the debate and the national significance of the project, Dublin Corporation as the planning decision-maker was left in a quandary regarding how to acquire the 'must-have' National Conference Centre, without having to grant permission for the controversial 557,418 square metre development. A representative of Treasury Holdings has since remarked that he believes the fundamental problem with the proposal was the sheer scale of the plan and supporting documentation, which highlighted the inability of Dublin Corporation to deal with applications of this size. In August 1999, full planning permission for the NCC and one adjacent office block was granted subject to more than fifty conditions. In addition, outline planning permission was provided for the rest of the project subject to a cap on development (including the NCC) of 418,063 square metres. The permission granted by the planning authorities highlighted a number of issues of concern, including:

- the dubious environmental quality of the layout;
- inadequate provision for a twin-track LUAS (light-rail) line to run to Guild Street;
- a requirement on the developers to contribute £2 million towards the construction costs of the new Macken Street Bridge;
- a requirement to provide more family-oriented housing with children's play areas, in order to achieve a sustainable and balanced community profile.

Not surprisingly, this decision proved unacceptable to everyone, and the developers argued that, by imposing such restrictions, the future of the National Conference Centre was thrown into serious doubt. An additional seven groups, including residents associations and local councillors, found the decision of the local authority to be fundamentally flawed. In a surprise move, Dermot Desmond, the original founding father of the IFSC, also lodged an objection to the planning appeals board. It could be argued that this in some way acted as a counterweight to the legal and financial power of the scheme proponents, but many local representatives disagree with this analysis. They have also expressed disappointment in the position adopted by the DDDA arguing that there was an expectation that the DDDA would 'fight for us'.

The oral hearing opened in February 2000 and was important in determining the future of the site, but it also raises more general questions regarding the democratic credentials of the planning system. Although all the appellants had in theory an equal chance to forward their arguments, it highlighted the inequality of opportunity available to ordinary people to participate in the planning process. Gerry Fay of the North Wall Community Residents Association highlighted the key issue that the hearing brought to light:

> It forces a small community like us, whose existence is threatened, to hold sponsored walks and bob-a-job days to pay for professional representation at this hearing. If this was a normal court, we would be provided with legal aid. We are going in against people with hundreds of millions of pounds behind them.
>
> (*Irish Times*, 21 February 2000.)

There is clearly a major gap in Ireland between the theory and practice of participation in the Irish planning system, and this was represented in media reports as an epic David *vs* Goliath style battle. Yet, although the intervention of Dermot Desmond dramatically altered the balance and significantly levelled the playing field, it did little to provide the local residents, those most affected by the development, with the power to influence local decision-making. Consultants employed by the financier argued strenuously against the development on a number of grounds. From an economic point of view, they argued that the development would cost the tax payer £525 million over 11 years as part of what was described as a rich menu for developers including a wide range of tax breaks and concessions (*Irish Times*, 27 February 2000). Architect Paul Keogh stated that the excessive height and mass of buildings was not aimed at creating a landmark building but rather was intended to create a hugely profitable commercial and residential development. Rather than following the European norm of six storeys, this development attempted to re-create Manhattan in Dublin. Michael Smith, chairman of An Taisce (National Heritage Trust), described the development as a 'squandered opportunity for perhaps the most exciting urban design brief in the history of the State' (*Irish Times*, 24 February 1999).

In contrast to the objections raised against the development, the project architect – Kevin Roche – defended the height of the buildings as a trade-off against the amount of open space that would be created. In language echoing the ideas of Le Corbusier that by increasing density one could maximize amenity space, Roche stated that the raised park at the centre of the scheme was only one element in the 50 per cent open space created across the entire site, to which there would be total public access. Countering accusations of being a 'height freak', Roche argued that although the buildings were only low-to-medium rise, he would not be averse to lowering the height of the buildings if required.

Perhaps in an attempt to deflect the arguments away from the core issues, the debate became highly personalized. Proponents of the project asserted that the rationale behind Dermot Desmond's objection was that the Spencer Dock development would have highlighted the inadequacies of the existing International Financial Services Centre complex, of which Desmond had been a key driver. Instead of simply focusing on constructing the most profitable elements of a scheme, as had occurred on the original Custom House Docks site, Treasury Holdings argued for developing Spencer Dock to illustrate how a real mixed-use development could function successfully. However, based on the tone of his comments to Government during discussions over the ecosphere project, mentioned earlier in this chapter, it may be more likely that Desmond recognized the failure of his original scheme to realize its mixed-use aspirations. He willingly raised this concern at the oral hearing, and described his intervention in the Spencer Dock debate as motivated by a desire to prevent the same thing happening again within docklands.

Following months of deliberation, the Appeals Board issued their findings in July 2000, ending almost two years of dispute and granting permission solely for the conference/exhibition centre and associated landscaping and site works. Permission was denied for all other elements – two hotels, nine office blocks and eleven apartment blocks – for a variety of reasons, namely that

> the proposed development, by reason of its scale, bulk, mass and campus-style layout, would constitute an inappropriate form of development for Dublin, would fail to integrate with the pattern of development in the city, would materially contravene the provisions of the current development plan for the area and would, therefore, be contrary to the proper planning and development of the area ... [and] it is considered that the proposed development of Phase 4 would result in a substandard form of development which would seriously injure the amenities of property in the vicinity and of future residents in the area.
> (An Bord Pleanála, Application 0600/99, Third Schedule.)

Local representative Tony Gregory described this as 'the most significant victory I've seen for any community group in the time I've been around and it was largely to do with their involvement in it'. In condemning how the development as a whole was handled, architect David McConnell has criticized how the conference centre became a political football and has argued that it 'should have been dealt with as an independent entity. The Government should have proceeded with it on the basis that it was a public good and necessary for the development of so many things connected with the city' (*Questions and Answers*, RTÉ television, 17 July 2000).

The politics of the Dublin waterfront

Similar to many flagship projects, some of which have already been discussed earlier, the Spencer Dock development in Dublin became a highly political rather than an urban planning or development project. Figure 133 summarizes

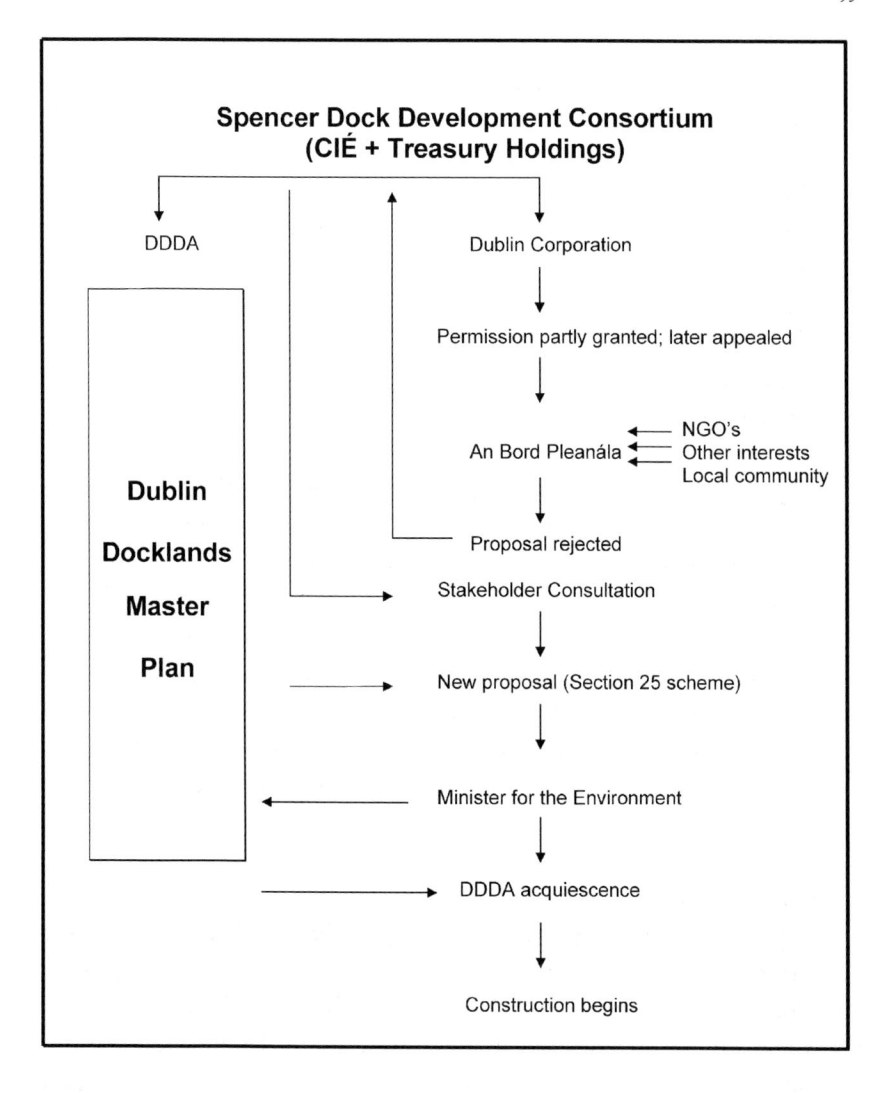

```
Spencer Dock Development Consortium
         (CIÉ + Treasury Holdings)

DDDA              Dublin Corporation
                        ↓
            Permission partly granted; later appealed
                        ↓
                                    ← NGO's
            An Bord Pleanála    ← Other interests
                                    ← Local community
                        ↓
Dublin      Proposal rejected
Docklands           ↓
Master      Stakeholder Consultation
Plan                ↓
            New proposal (Section 25 scheme)
                        ↓
            Minister for the Environment
                        ↓
            DDDA acquiescence
                        ↓
            Construction begins
```

133 The development process at Spencer Dock. (N. Moore.)

the path to development undertaken in this area, highlighting the complexity of the institutional landscape and the lack of integration between various agencies. In many ways mirroring the problems and issues faced in other cities such as London or Cardiff regarding the politicization of planning, the Spencer Dock project also has a number of unique attributes, most notably the speculative gamble by the developers to direct their planning application through regular channels rather than fast-track the development through the

DDDA in 1997/8. This is probably easily explained by the poor relationship between, and intransigence of, the two parties involved. The decision to use the regular planning route may have been an attempt by the Spencer Dock Development Company (SDDC) to maximize disproportionately their return on investment. In their proposal for 557,418 square metres of commercial space to subsidize the construction of the National Conference Centre, the SDDC seems to have tried to overwhelm the planning authorities, push the development close to the time limit for incentives imposed by the European Union and thus force the authorities to acquiesce to their proposal.

The Spencer Dock saga also highlights the dramatic impact and effects of government intervention in the urban planning and development process. In the early 1980s, following the formulation of the 'Gregory Deal' and the decision to nationalize the 27-acre landbank at the Custom House Docks, proposals for development sponsored by the Dublin Port and Docks Board that had been progressing through the planning system, were halted. The uncertainty heralded by government acquisition of the site destroyed any hopes of private sector investment or indeed interest in the area. Ironically, in direct contrast with, and some might argue because of, what had occurred a few years earlier, the period from 1986 onwards was characterized by intervention in order to progress and speed-up development (as the previous chapters have discussed in detail). The lessons learnt from 1986 onwards clearly illustrate that government involvement and close monitoring of development is an imperative in guaranteeing relatively balanced regeneration. If we were to try to conceptualize government intervention, we might argue that it occurs in a cyclical fashion. A direct negative correlation is apparent between the health of the land market and the role of the Government. When the property market is weak, governments intervene to streamline the process and encourage development with a range of fiscal and other incentives; when the property market is strong and rampant speculation becomes evident, governments must intervene to regulate and control the private sector. In the case of the Spencer Dock project, the intervention of the State, represented by the DDDA, became imperative in ensuring that corporate considerations and profit-making did not solely determine the future of this urban quarter. In hindsight, one wonders why other State agencies could not have stepped into the debate over Stack A in a similar way to ensure that the original commitment to a museum facility was upheld.

The emphasis on the city as a physical commodity to be bought, sold and redeveloped for profit has made it increasingly difficult for planners to

134 The new Liffey vista, north quays, 2005. (J. Brady.)

guarantee the achievement of social regeneration and to safeguard a viable future for local communities, and the lack of interest shown by many urban development corporations, not just in Ireland, in engaging in such activities has been the subject of much criticism (Merrifield, 1993). Separate development agencies, outside the remit of the local planning authority, have tended to be more developer or private-sector led than socially-oriented. The standard argument for retaining development control with local authorities, in this case Dublin Corporation, is that the planning and development process is designed to achieve the 'common good' rather than speculative economic gain. Ironically, in the case of the Spencer Dock scheme, it would appear that the local authority was so overwhelmed by the size of the planning application that it received and by the national significance of the project, that it was paralyzed as to the most appropriate decision to meet the 'common good'. Local residents were also disadvantaged in terms of their ability to influence the decision-making process, because of a lack of both time and skills to negotiate such a complicated and voluminous set of documents. Had the developers pursued their development through the non-elected body, the DDDA rather than the local authority, community representatives may have paradoxically had greater influence over the development proposal through their representation on the Council of that Authority. The complexity with which large-scale flagship projects are developed and implemented is very clearly exemplified by this particular project. On 5 April 2007, the Government finally awarded a contract for the construction of the National Conference Centre and committed itself to a site at Spencer Dock.

Dublin's new city quarter: Spencer Dock today

Following the decision of An Bord Pleanála to reject the initial plans for construction at Spencer Dock, the Dublin Docklands Development Authority published a draft Section 25 planning scheme for 12.5 hectares of the site, comprising the area south of the Sheriff Street Bridge. Having already lost a number of years of potential development time, Treasury Holdings responded rapidly to this development and submitted a much more street-based and European-style proposal, as originally desired by the DDDA. In January 2003, the development consortium was awarded a Section 25 certificate for four hectares, effectively a licence to begin development. The new scheme will have a significant residential element, comprising a series of eight and nine-storey apartment blocks with a range of different unit sizes grouped together by the canal, and a number of duplex units fronting on to the canal with generous gardens. The desirability of this prime docklands location was reflected in the dramatic price increase between the first and second release of apartments to the market. In 2003 when the first phase was launched from plans, one-bed apartments were priced from €300,000, while in March 2004 when a further 75 units were released this had increased to €375,000. The first phase is due for completion before Christmas 2007 and under the requirements of the Master Plan will provide 99 affordable and 106 social units out of a total of 600 units. Not only is a range of new development being accommodated, including housing, a community crèche, and a headquarters building for PriceWaterhouse Coopers, but older historic structures are being retained including the Northwestern Railway Hotel and Woolstore (Figure 135).

While a 1.62-hectare site was reserved for the National Conference Centre within the Spencer Dock boundaries in the new submission in 2001, the Government only gave 'preferred bidder' status for this element of the scheme to the development consortium in late November 2005, having launched yet another competition for the project. The Office of Public Works has undertaken negotiations with the development consortium in a bid to deliver the project through a public-private partnership (PPP) and finalized the contract in April 2007. Given the make-up of the development consortium

135 Northwestern Railway Hotel, early-twentieth century. (Postcard.)

136 Northwestern Railway Hotel in 2005. (J. Brady.)

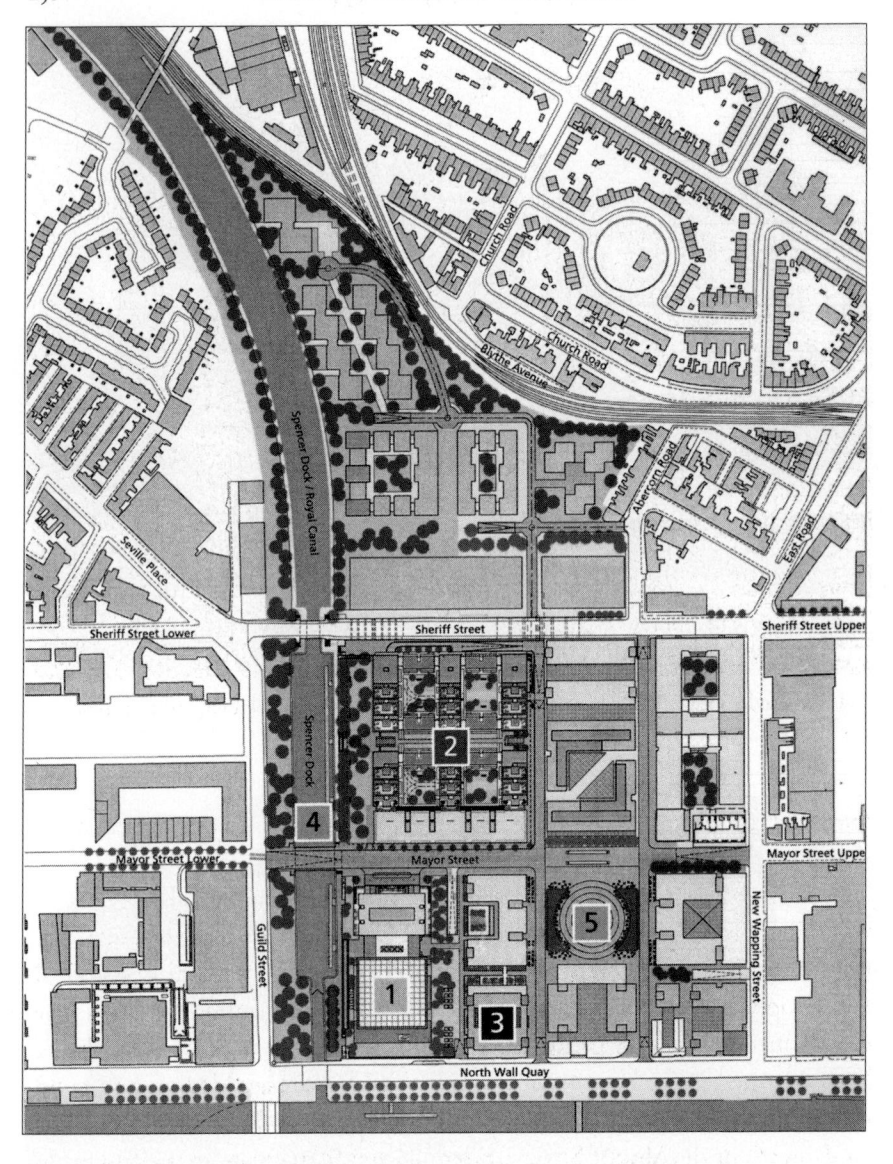

137 Revised Spencer Dock development. (Advertising Brochure.)
1. National Conference Centre, 4. Linear Park, 5. Performance Square.

(itself a PPP), the State is now in an unusual position, having a role both in the public and in part of the private sector (through CIÉ) interest. Construction is due to begin in 2007 with an anticipated completion date of 2011. The office buildings that surround the conference centre are complete and the first

138 Revised Spencer Dock scheme and old Northwestern Railway Hotel, 2005. (J. Brady.)

tenants, PriceWaterhouse Coopers and Fortis Bank, have occupied them since Easter 2007.

Although the overall scheme currently under construction was a far more welcome proposal than the original plan for 557,418 square metres in high-rise buildings, the previous chapter highlighted one of the remaining bones of contention between the developers and the local community over the clustering of social and affordable housing in one corner of the overall site. Perhaps some lessons have been learnt, as this has been avoided in the second part of the development, outside the Section 25 area and focused on the CIÉ lands north of the Sheriff Street Bridge. Rather than create more dissention and controversy, the CIÉ/Treasury Holdings consortium has demonstrated a significant shift in attitude and become the first developer, outside the Ballymun Regeneration Project, to utilize a community planning approach to development. This may have been driven by the fact that, as this second phase is outside the remit of the DDDA, the developers had to satisfy the city council that adequate public consultation had been undertaken. In June 2003, a British firm experienced in this field was invited to docklands to run a series

139 Aerial view of the Spencer Dock scheme, 2005. (DDDA.)

140 Selling the Spencer Dock scheme, 2006. (J. Brady.)

141 Selling the Spencer Dock scheme, 2006. (J. Brady.)

142 Spencer Dock scheme, summer 2007. (J. Brady.)

of public meetings to harness the ideas of the general public for future development of the 2.23-hectare site. Rob Ticknell explained that:

> Everyone in the area was quite tired of the standard public consultation process, which had yielded nothing much for them, so we decided to take a completely new approach of working with the local community and involving them directly in designing the scheme.
>
> (*Irish Times*, 3 July 2003.)

The key requirements identified by the local community were to create new links between the East and North Wall communities, which have for many years been separated by the railway, and to improve the overall permeability and openness of the whole area. The latest plans envisage a pedestrian bridge over the railway line, the demolition of high stone walls which act as a barrier to movement, and the development of a one kilometre public park along the canal which has been a key priority for residents. A new event space for markets, fairs and festivals is proposed for the dockside where people can freely congregate and interact, while a 'green' building will be constructed to terminate the site at the northern edge. This approach and outcome are a significant improvement on the manner in which the original plans for the whole district were conceived, but there is still some lack of trust between the stakeholders: there is much more communication and co-operation occurring at present in comparison to the situation three years ago, but there is still a cynicism on the part of the developers regarding the motivating factors of the local community:

> They may or may not start marrying each other. But with the Spencer Dock development there will be a whole new raft of people moving in and with them will come employment and cold hard cash. That should cheer up the local community no end.
> (*Irish Times*, 21 November 2003.)

Environment and development

The two biggest drivers of controversy in docklands so far have focused on the content of development proposals and the nature of decision-making, yet more recently and indeed in the future, the biggest challenges for the Dublin Docklands Development Authority will be to address the problems of redevelopment on more difficult or 'problem' sites. In cities throughout the developed world but particularly in Europe, the so-called rustbelt of the north eastern USA, and Ontario, Canada, increased attention is being paid to areas that have become abandoned because former industries have either closed down or moved elsewhere. This has occurred in response to global economic restructuring and the increasing movement of manufacturing to emerging cities in Asia. Since the 1970s, many cities in the more developed economies have had to discover new roles and attract the new service-based industries. Waterfront areas have been among the most challenging districts to regenerate, because of their history of transhipment and manufacturing of gas, chemicals and other noxious substances. The EU Expert Group on the Urban Environment has identified these contaminated sites or 'brownfields' as not just an ecological but also an economic and social problem; but, they also argue that brownfields provide many opportunities for urban consolidation, for densification, and for new economic activities as well as for supporting 'nature conservation, biodiversity and climate protection' (EU Expert Group, 2001, p. 13).

This potential has been recognized by the DDDA who identified two major brownfield sites with acute environmental problems in the 1997 Master

143 Gas production, Grand Canal Docks, 1983. (Crampton's Photo Archive.)

144 Grand Canal Docks derelict site, 1994. (J. Brady.)

Plan, one being the former gasworks at the Grand Canal Docks (Figure 143) and the second being the Poolbeg Peninsula. The very different contemporary land uses in these areas, have resulted in a need for different responses; and whereas the Grand Canal area is now undergoing complete reconstruction, the Poolbeg Area is still awaiting a decision on the optimum kind of future development and is subject to growing dissatisfaction among a range of interested parties.

In recent years, in response to the rapid and unprecedented urbanization that has emerged from the Celtic Tiger economic boom, urban management has become a critical issue in Ireland. The National Spatial Strategy 2002–2020, *People, Places and Potential*, argues for a strengthening of existing policies of counter-urbanization to halt the further development of a large urban area on the eastern seaboard. For this to occur, a policy of containment or consolidation, focused on land recycling in urban areas is needed. The overall objectives are to implement effective land use and public transport policies, and, for the first time in Irish urban policy, the potential for brownfields to contribute to intensification and compaction has been made explicit. The document argues that within the Metropolitan area 'a systematic and comprehensive audit of all vacant, derelict and underused land [should be undertaken] to establish its capacity to accommodate housing and other suitable uses'. It stresses that 'where appropriate, local authorities should be pro-active in using their existing powers (such as those under the Derelict Sites Act, 1991) to facilitate the assembly of fragmented sites and to encourage the relocation to more suitable sites where there is inappropriate land use within city/district centres', and that 'intensification can be achieved without compromising amenity'. The document further encourages local authorities to 'examine the potential of declining industrial and warehouse estates for re-development for new economic activity' (para. 3.3.1.a). A practical prescription to ensure consolidation of the Greater Dublin Area and other large urban centres within Ireland is given: 'the efficient use of land by consolidating existing settlements, focusing in particular on development capacity within central urban areas through re-use of under-utilized land and buildings as a priority'. This suggests that brownfield regeneration has the potential to become a cornerstone of future spatial development, and leading the field in this area has been the Dublin Docklands Development Authority. The authority has tackled and indeed overcome the development constraints of one of the most strategically sited brownfields in Ireland, at the Grand Canal Dock.

145 Grand Canal Docks derelict site, 1994. (J. Brady.)

146 Grand Canal Docks derelict site, 1994. (J. Brady.)

Unlike much of the docklands area, which is relatively cut off from the rest of the city, the Grand Canal Dock is highly accessible by public and private transportation and is familiar to anyone crossing the city regularly by car or train. For many years, the gasometers that dotted the skyline in this area could be spotted from quite a distance, setting this area within the city apart from any other. The development anticipated by the Dublin Docklands Development Authority for this area, and currently under construction, closely resembles the original 1987 Planning Scheme for the Custom House Docks Area, as it aspires to the creation of a vibrant mixed-use development with significant residential land use, comprising dwellings, cultural, hotel, local retail uses, and commercial functions. Because of the slow-down in the office market and the need to breed vibrancy and promote the 'people factor' in this district, the balance of the project was rapidly altered with the greater emphasis (60% of total build) on residential uses, creating an entirely new living quarter in an area of the city that has always been associated with industrial activity. Even so, redevelopment has not been straightforward, and the development authority have had to deal with the legacy of industrial contamination and ensure full remediation occurred before they could generate any significant private sector interest in the area.

Remediation and redevelopment at the Grand Canal Dock

Until the late 1970s, the Grand Canal Dock Area was synonymous in Dublin City with the production of town gas. This history resulted in a need to undertake a comprehensive risk-assessment exercise before any plans for regeneration were made in this zone. In their assessment, Parkman Consultants identified a range of chemicals within the soil, all either products or by-products of gas manufacturing including cresol, toluene, xylene, ethylbenzene, benzene, ammonia and naphthalene. The presence of these chemicals and the lack of any desire to take on the risk of a contaminated site may explain why the site lay derelict for so long at what is a prime urban location. The biggest barrier to regeneration and perhaps the greatest concern from the developers' perspective is the lack of clarity regarding liability for derelict and particularly contaminated sites in Ireland. Internationally this has been a major stumbling block to regeneration, for it leaves unanswered a number of questions, including who pays for the clean up of contaminated sites, how clean is clean, and who is responsible for these sites once they have been de-contaminated. Whether the polluter or new owner bears responsibility for any future liability from historical contamination is a critical issue, and in Ireland (where the national media and politicians are increasingly referring to the emergence of a litigation culture) this is a particularly salient issue.

Apart from the end-user taking a claim against a particular developer, there is also the employer liability that may arise from the remediation of a contaminated site. There are also Constitutional difficulties, as the Oireachtas traditionally does not declare an Act to be an infringement of law retrospectively (McIntyre, 2003). Therefore, if a landowner created a brownfield but the action was not illegal at the time, then he or she bears no responsibility for the contamination. As it is virtually impossible to assign legally liability for historically contaminated land in Ireland, 'responsibility [for regulating liability] is ultimately likely to pass to the public sector, perhaps explaining the Irish authorities' lack of resolve to tackle this problem' (McIntyre, 2003, p. 117).

These issues and the delays in redevelopment that they generate are not specific to Dublin. In a range of major cities around the world from Pittsburgh to London and Sydney, similar problems are emerging albeit on different scales. In the Canadian capital, Ottawa, the area known as LeBreton Flats which is located just 1.6 kilometres south west of the city centre, has displayed similar problems to the Grand Canal Dock in Dublin. Overlooked by Parliament Hill, this area has remained undeveloped since 1965. For forty years 26.7 hectares remained derelict at the heart of the city, the key reason being that industrial activities had produced a significant amount of contaminated land which nobody could agree how to treat. In the late-1990s, two consultancy reports were commissioned to determine how to handle this problem. The first, by Aqua Terre Solutions, recommended in March 2000 that the Government should transport the polluted soil to an offsite landfill; while the second, by IBI Group, proposed that the polluted soil should be dug up and reburied in the immediate vicinity rather than hauled off to a distant dump. They argued that the ideal would be to dump it on the former Nepean Bay landfill, on the bank of the Ottawa River at the western edge of the LeBreton Flats, saving the Government $20 million in cleanup costs and making proposed development more affordable.

The problem with this second proposal was that it overtly highlighted the short-term thinking that permeates a lot of urban development and many planning organizations, trying to push the problem somewhere else, in this case into the adjacent river, or into the future. At the Grand Canal Docks in Dublin, this kind of approach was not even considered, and Peter Coyne, then CEO of the Dublin Docklands Development Authority, summed up

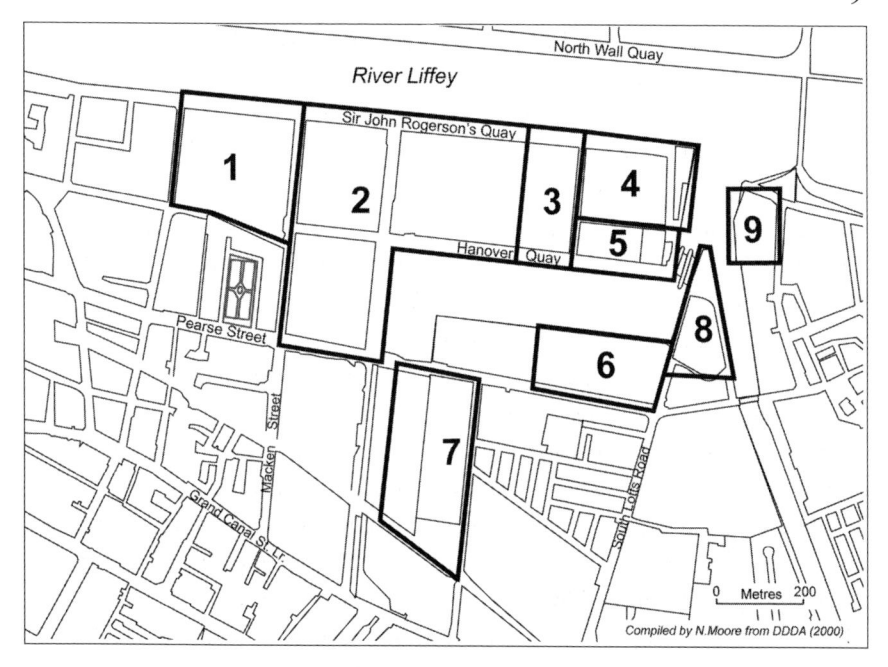

147 Original Grand Canal Docks development scheme. (N. Moore.)
1: 5/6 storey commercial and residential; **2**: Former BGE site; **3**: Commercial; **4**: High landmark building, retail and bar use; **5**: Cultural/retail, restaurant or bar use; **6**: Residential; **7**: Residential and tall landmark building; **8**: Infilled graving docks and Commercial facility with maritime theme; **9**: Proposed residential.

their attitude as a belief that 'you must spend money to make it'. Extensive remediation was undertaken, and the eventual bill for the decontamination of the 10.1-hectare site at Grand Canal Harbour has amounted to €50 million. For construction to begin, the site had to be certified by the Environmental Protection Agency, which played a key monitoring role during the two-phase remediation process. This agency licensed two separate development sites (both of which abut Sir John Rogerson's Quay within the Grand Canal Harbour boundaries) in October 1999. The first part of remediation involved the construction of an 8m deep and 2km long bentonite wall into the soil to ensure that any contamination on the site or any dislocation caused by remediation would not pollute adjacent areas. The entire site was then excavated to a depth of four metres, and the groundwater, which may have been contaminated, was pumped out to a sewer via treatment works. Hazardous soils were shipped off-site to a number of EU countries, including Belgium, where the soil was burnt to clean it and then utilized in the construction of

148 Aerial view of Grand Canal environs, 1997. (DDDA.)

flood defences. Most of the soil was treated on site, as the EPA prefer, and then reused. The waste licence for the first part of the site was surrendered in December 2001 and that for the second part a year later, following the removal of over 134,000 tonnes of material and the award of a clean bill of health by the EPA.

This achievement was a key victory for the DDDA, who had demonstrated that such risks could be overcome, and it bodes well for future development in the country as a whole, given that the EPA estimates there to be about 2,400 brownfield sites in Ireland, although none as large as this one. Beyond the physical remediation, one of the challenges most difficult to address during the whole process was public perception. In an important public relations exercise and to address and curb community concerns, fortnightly public meetings were held by the DDDA to report on progress. Now the

149 Rapid pace of development, 2005. (J. Brady.)

authority argue that the air around this site is among the cleanest in the city, and is certainly cleaner than College Green, which has some of the largest pedestrian flows in the city. An additional investment of €7 million for infrastructure and other environmental improvements was made by the DDDA, but this has been easily justified by the fact that the authority has already generated over €200m in land sales from this site.

A range of tenants has been secured for Grand Canal Harbour, the 10.1-hectare site being developed directly by the DDDA within the overall 37.2-hectare dock area. These include the headquarters of legal firm McCann Fitzgerald, who will occupy Riverside 1, adjacent to the Ferryman hotel. This site was originally designated for the software firm, Novell, who proposed to pay €25.4m for the site and construct a 6/7-storey tower due for completion in 2002. The collapse of the technological sector in recent years precluded this occurring, and McCann Fitzgerald acquired the site for a bargain €12.5 million. When the building is complete, they will relocate from Harbourmaster 2 within the IFSC, but must find another tenant to take up the lease there as it does not expire until 2026. This marks a worrying trend in relation to the overall docklands area. Given its size and the likelihood that newer and better office facilities will continue to be constructed over the coming ten years, a

150 Aerial view of development, January 2005. (DDDA.)

151 View from the south, May 2005. (DDDA.)

detrimental impact may be felt in the original Custom House Docks site, where tax incentives have now ended. If firms identify a preferable location elsewhere in docklands, the prestigious nature of the original IFSC buildings may be undermined.

As well as the commercial element, which is after all not the majority land use within the Grand Canal Dock, the strength of the 'market' for a range of other uses appears to be strong, and the Park Hyatt Hotel Group have just recently confirmed that they will be constructing a 5-star hotel at Misery Hill and an adjacent Hyatt Residence Hotel. By the time the development is complete, there will be over 3,500 new apartments in the area and over 408,773 square metres of office space. Five 20-storey landmark towers will punctuate the site, the most well-known being the new U2 tower which is under construction and will house a recording studio for that band. The whole development is anchored by a 0.6-hectare piazza, Grand Canal Square designed by Martha Schwarz, adjacent to which will be a performing arts centre, designed by the internationally renowned architect, Daniel Libeskind, and developed by Heritage Properties. In contrast to the attitude adopted towards Stack A and similar to the incentive used to attract the National College of Ireland to docklands, part of the site for this new cultural amenity

152 Pearse Square, Pearse Street, overshadowed by new development, 2005. (J. Brady.)

has been provided free of charge. A development levy applied to all commercial tenants should raise approximately €500,000 to subsidize the centre, which should attract a wide range of visitors to the site outside the traditional working hours. However, this has not been developed without difficulty; the Government announced in May 2004 that development would start soon, yet it is unlikely to open before 2008. Joe O'Reilly, one of the developers behind the Dundrum Town Centre, will have a major stake in the centre when it is complete.

Leaving aside the delay in this particular part of the project, what the DDDA has succeeded in doing is engineering a dramatic transformation in the desirability of this former marginal part of the city centre, and the result has been massive spin-off benefits for landholders in the immediate vicinity. New projects outside the Grand Canal Harbour boundaries include the Duffy Lally office development at South Bank Quay and the Ellier Developments project at Hanover Quay. Treasury Holdings have just completed contracts on 21,368 square metres of office space fronting Sir John Rogerson's Quay, while another joint venture development with CIÉ will create a mixed office-apartment scheme near the Grand Canal Dock DART station. If the mixed-use development is successful, this may prove to be a significant turning point in the perception of brownfield areas nationally. It may also signal a new trend in Irish planning and

153 The old chimney amidst new development, 2005. (J. Brady.)

development away from greenfield sites and back into these residual, city-centre 'problem areas', further contributing to the national spatial strategy goal of consolidating the Dublin city region. The Grand Canal Dock project illustrates that where there is a will and central government support to achieve a goal, it is possible to balance environmental and development objectives and can in fact result in economic and social returns far beyond the initial investment.

Future development: the contested nature of the Poolbeg Peninsula

Although different challenges have faced developers, policymakers and communities in each of the three project areas already examined, they all have

154 Millennium Tower: High-rise apartment living, 2006. (J. Brady.)

155 Inner Grand Canal Quay prior to its redevelopment in 2006. (J. Brady.)

156 Grand Canal Square takes shape, summer 2007. (J. Brady.)

157 The new streetscapes of docklands, 2007. (J. Brady.)

158 New commercial development at South Bank Quay. (J. Brady.)

some characteristics in common, including a residential and commercial element. The Poolbeg Peninsula is very different and will in the near future provide even greater challenges for regeneration than those already encountered. If most Dubliners were asked to describe the area, they would probably identify the two ESB chimneys that dominate the landscape as emblematic of the area (Figure 159). Others who have ventured beyond the roundabout leading to the East Link toll bridge might describe this as the engine of the city. The existence of scrapyards, the two electricity stations and the new sewage treatment plant commissioned in 2003 would be cited in support of this description. Because the new Combined Cycle Gas Turbine (CCGS) power plant at Ringsend and the gas/oil power station at Poolbeg generate the equivalent of 35 per cent of national electricity, this analogy is fairly accurate. While most of the major metropolitan utilities are located in this district, they co-exist quite unusually with natural amenity areas such as the Irishtown Nature Reserve. While the docklands authority recognizes that there is huge potential for change in the district and that 'underutilization of land remains a problem on the Peninsula', it also concedes that 'utility functions on the Poolbeg Peninsula have been reinforced with the construction of two new power stations and the waste water treatment plant' (DDDA, 2003, p. 16).

159 Aerial view of the Poolbeg Peninsula. (J. Brady.)

160 The industrial landscape of the Poolbeg Peninsula. (DDDA.)

161 Local authority housing on the south-western edge of the Poolbeg Peninsula on reclaimed land. (J. Brady.)

One of the general objectives of the authority is to develop, in co-operation with Trinity College Dublin, a Technopole, and the southern part of the Poolbeg Peninsula is considered an appropriate location as it would also act as a buffer zone between residential development and industrial zoning. But as well as significant industrial and utilities activity, the peninsula has a unique natural environment and amenity potential which it is proposed to harness for the benefit of both local communities and visitors. Policies to direct the development of the area focus on the generation of local community employment through the Technopole concept; the possible development of an industrial heritage area near the Old Pigeon House Hotel; and the improvement of the physical environment through the protection of defined natural habitat areas and the development of an eco-park to reinforce the existing nature park, to provide landscaping particularly between the utilities and the coastal parkland area, and to provide a feeding ground for Brent geese.

There are undoubtedly significant obstacles to development in this area, but it also provides excellent opportunities, given that large tracts of land remain in public ownership through bodies like Dublin port and the ESB, which should at least theoretically make comprehensive redevelopment easier.

Having said that, CIÉ's ownership of the rail yards at Spencer Dock did not make development any quicker or easier at that location, and, if anything, it probably complicated matters. The sale of the Irish Glass Bottle Company site opened up a very large site for redevelopment and, to accommodate this change, the most recent Master Plan altered the land-use zoning from 'industrial uses and the provision of local employment opportunities' to 'mixed use including residential, to allow high quality development front onto Sean Moore Park and Dublin Bay'. The 2003 Dublin docklands Master Plan recommended extending the proposed Section 25 status on the southern edge of the Poolbeg to include the IGB site, but the authority suggested that this would be a longer-term objective because it may only happen 'towards the end of Phase 2 of the Master Plan period' (DDDA, 2003, p. 92). Nonetheless, in December 2006, the DDDA announced plans to invest €32.17 million in a joint venture that would buy the site from South Wharf plc. This 10.1-hectare site was bought by developer Bernard McNamara in partnership with the Dublin Docklands Development Authority and is likely to facilitate a range of future uses that will ultimately change the character of the area forever. While the potential of the site is recognized by everyone, Frances Corr from the Combined Residents Association in the area has identified a number of critical issues of concern. She points out:

> We don't want to lose the three villages' atmosphere by putting a Manhattan on the corner. The character of Ringsend, Irishtown and Sandymount could be endangered by the development of the Glass Bottle site. There's a need for a supermarket, retail, DIY and perhaps a cinema in the area. Everything I'm looking for, I have to go somewhere off the M50 to find. This would be more in keeping with what people want. The site is contaminated and rubbish will have to be taken out and probably put into the proposed incinerator site. And who's going to pay top dollar to live beside an incinerator? Families won't take the risk. It's also a flood plain area. Local submissions to the National Plan for the docklands requested a mixed use of social and affordable housing and they didn't wish to see high-rise buildings of more than six to eight storeys. There has to be a live, vibrant mix.
> (*Irish Times*, 2 November 2006.)

Accessibility will also be a critical issue for any development that will take place, given the present heavy traffic flows in the area, and the single route in and out of the peninsula presents a major difficulty. Another limiting factor, from a construction perspective, may be the stability of the site, since it is mainly reclaimed land and may be incapable of supporting large-scale development. Taking into account the historic and contemporary land uses, there is also a significant possibility that contamination will prove to be a problem at this location as it did at the Grand Canal Dock. Complicating matters further is the institutional complexity in the ownership. While the area falls within the remit of the DDDA, they are not the planning authority in the area. Dublin City Council remain the key driver or constrainer of development at this area and have recently commissioned a strategic development framework report for the area that they are beginning to market as 'Dublin South Bank'.

The draft Poolbeg Framework Plan was completed in October 2003, but the City Council delayed publication until after the local elections in summer 2004. The primary reason for approaching the publication of the plan in this way may have been because of the debate surrounding a Waste to Energy Facility proposed under the Dublin Waste Management Plan (1998). This incinerator would process waste from all four Dublin local authorities and has been the subject of massive opposition from local politicians and residents in the area. Indeed, one of the candidates in the June 2004 local elections ran specifically on an anti-incineration platform, highlighting the political nature of the project. During the general election campaign of summer 2007, this issue again reared its head, and the future of the Poolbeg Peninsula is becoming a critical political as well as an environmental issue. Residents and local political representatives argue that when the plan was originally published it was unrealistic, identifying the district as the potential home for 18,000 people without adequately addressing issues such as traffic and transport management, soil contamination and other infrastructural issues. They also claimed that it disregarded the DDDA Master Plan which should have informed any proposals developed for the zone as well as pre-empting the new City Development Plan. However, the hopes of local residents to influence and discourage the development of an incinerator in this area were dashed when the Dublin City Development Plan, 2005–2011 identified particular industrial zoning for the proposed incinerator site. In a clear example of the questionable democratic credentials of our local authority structures, the elected members of the Council or city councillors specifically opposed the siting of the incinerator on the Poolbeg Peninsula at a meeting on 14 November 2005, yet this was subsequently overruled by the City manager. Dublin City Council along with the other Dublin Local Authorities, Dun Laoghaire Rathdown, Fingal and South Dublin County councils have applied to An Bord Pleanála for approval on the Poolbeg Peninsula of an incinerator that will burn 600,000 tonnes of household, commercial and industrial waste per annum. Yet, in response, local activists gathered over 3,000 objections against the project. Almost all political parties have indicated their opposition to this proposal, not surprising given the strength of feeling among the local community. John Gormley of the Green Party, now newly-appointed Minister for the Environment, has previously argued that:

> The proposed incinerator would impact negatively on the potential development of the area. As the Green Party predicted, this plant is set to burn 760,000 tonnes of waste per annum, making it one of the largest mass burn incinerators in Europe ... If it goes ahead, the traffic to and from the incinerator would make living in the area very unpleasant.
> (Green Party press release, 20 October 2006.)

In January 2007, former Fianna Fáil councillor and recently elected TD, Chris Andrews, brought the case of the incinerator to the European Commission, arguing that the Government did not give proper consideration to a European directive that requires it to ensure waste is disposed of without endangering human health or the environment. His argument centres on the lack of plans to deal adequately with the impact of the proposed incinerator on surrounding communities and highlights local experience with previous large-scale infrastructure projects as adding to the concerns. In contrast to the widespread marketing and assurances made at the time, he believes that the 'Dublin waste water treatment plant, which was opened on the Poolbeg peninsula in 2004 and was promised to be run at the highest of environmental and safety standards has proven to have severe health and environmental implications' (*Irish Times*, 30 January 2007).

In the context of the composition of the new national government and the proposals for development that are beginning to gain increased attention for this area, including the proposals of the Progressive Democrats to develop a mini-Manhattan on the Poolbeg Peninsula, the siting of an incinerator at this location appears to be losing momentum. The role and operations of various agencies in the area are also coming under increasing scrutiny and are subject to the kind of questioning that shows the difficulties and challenges in regenerating such a large area with such unique attributes. For example, the democratic deficit inherent in the way in which the siting of the incinerator has been forced through by an unelected City Manager in opposition to the desires of the elected members questions the structures on which the entire local democratic process in Ireland is built. The role of the DDDA in entering a joint venture with a private developer on the Irish Glass Bottle site has also raised significant questions that have ramifications far beyond the city itself. John McManus writing in the *Irish Times* on 4 December 2006 questioned whether it is

> really the business of a State development agency to be getting involved in such a massively leveraged project? What it brings to the party is the statutory power to have the developments exempted from planning, which was granted to it by the Government in order to expedite its mandate to redevelop Dublin's docks. A fair question is: what will happen if the project runs into difficulties and McNamara's solution is a material change to the current proposal, such as adding a few extra floors to the apartment blocks and scrapping the affordable housing? The DDDA will then find itself in the position of either waving goodbye to the taxpayers €32.17 million or going along with something that is potentially contrary to its own mandate.

Questioning development?

Although each of the developments outlined in this chapter have raised significant questions in relation to the current and future development of part of our capital city, the themes that they address have much broader implications. How we as an economically booming nation deal with our heritage will continue to be an important issue. Whether we recognize the value of particular areas and sites in more than economic terms is an issue on which the jury is still quite clearly out. While Stack A is now a relatively successful leisure destination for some, it has failed to meet its possibilities in terms of educating and engaging future generations with a significant era in the history of the city. This lack of awareness has also played itself out in Spencer Dock, where it appeared for a long time that economic imperatives would become the driving force, ignoring broader environmental and social considerations. One of the key outcomes of this controversy was the emergence of a more active and engaged citizenry willing to defend what they viewed as an attack on their area and history. A similar situation is emerging in the south docklands where many local groups are now actively contesting the proposed siting of the municipal waste-to-energy facility.

What has also become clear is the manner in which dramatic transformations can be wrought in an area when the appropriate institutional framework is put in place. While the Grand Canal Dock was once considered a marginal and problem area within the city, it is now proving to be one of the most desirable locations in which to reside. The ability of the State to intervene to create the appropriate conditions for development was critical to ensuring this success. Whether a similar approach will be taken in the Poolbeg Peninsula remains to be seen, yet the close linkages with the private sector particularly at the Irish Glass Bottle site may create more difficulties than provide solutions. Whatever the outcome, the major challenges over the next decade will be to accommodate a range of uses and users within close proximity of each other in this area, to maximize the efficiency of under-utilized sites and transform the Poolbeg Peninsula from the city's engine room to a paragon of sustainable urban development for the twenty-first century.

Continuing evolution

> We will develop Dublin Docklands into a world-class city quarter – one in which the whole community enjoys the highest standards of access to education, employment, housing and social amenity and which delivers a major contribution to the social and economic prosperity of Dublin and the whole of Ireland.
>
> (DDDA, 2002.)

Dublin docklands today has become a cornerstone of the city that embodies the rapid restructuring that has been experienced in Ireland over the last twenty years. Large tracts of the inner city have been transformed from derelict and depressed environments into vibrant and successful parts of the urban core. Dublin is unique among European cities in the exceptional levels of population and economic growth that it has recently experienced. While other cities, particularly in the more traditional European core economies of Germany and France, are coping with the phenomenon of 'shrinking cities', Dublin and its hinterland are struggling to manage dramatic expansion brought about in large part by the attraction of foreign investment to the city. The Globalization and World Cities research centre at the University of Loughborough have identified Dublin as an increasingly important urban region in global terms and a city that shows strong evidence of becoming a 'world city' (Beaverstock *et al.*, 1999). Peter Taylor (2003) in a comparative study of cities has ranked Dublin as the fifth most important European city in relation to its network or gateway power, and the twelfth most important network city in global terms. This is a measure of how embedded the city is in international economic activity, and indicators used include the number of international banking firms present, quantities of international investment funds located in the city, and the number of global firms with headquarters functions here. Dublin is now ranked ahead of other cities including Milan, Barcelona and Madrid as a significant gateway city in global economic webs of activity. The development of this profile has been largely facilitated by the growth of International Financial Services activities, the most significant proportion of which take place in the docklands. The redevelopment of this large urban quarter has become a central element of the 'Celtic Tiger' success associated with turn-of-the-millennium Ireland, and although the docklands project initially emerged out of the search for a solution to the inner-city crisis, redevelopment has become increasingly interconnected with the demands of the national and global economies.

The challenges of redevelopment

Although the Dublin docklands regeneration can be rated in general terms as a significant success, the evolution of the project has been fraught with difficulties and tension, in particular between the many different groups that have a stake in the future of the area. Initial difficulties focused on the manner in which development could be kick-started in what was widely perceived as a problem zone. The early plans of the Dublin Port and Docks Board were rejected by a range of agencies and groups in the early 1970s and again in 1980 and were condemned as representing an attempt to develop opportunistically an area without adequate consideration for the surrounding communities and districts. Yet a very different kind of opportunism became evident just a few years later as the area became embroiled in the controversial Gregory Deal. In the context of Coalition governments, there is nothing surprising or unusual about the kind of agreement reached between Gregory and Haughey at the time, and comparisons might be made with the ability of Independent TDs such as Finian McGrath to agree a similar deal with Bertie Ahern after the 2007 general election. The decision in 1982 to nationalize the Custom House Docks as part of the Gregory Deal was criticized by the Dublin Port and Docks Board, who clashed with central government over the inadequate compensation that they received when the site was compulsorily taken from them.

The tension between central government and various stakeholders within docklands during the early years of the project was not unique and has characterized the path to development in this area to the present day. Major tensions emerged in the early 1990s between the Custom House Docks Development Authority and community groups. Local residents were legitimately concerned with the lack of formal consultation and community engagement in relation to the direction of redevelopment. The mistrust generated at that time has only recently begun to be overcome through the formal consultation procedures required under the legislation introduced in 1997 to set up the Dublin Docklands Development Authority, but tensions still emerge in relation to particular projects and individual elements within them.

As with all major schemes in rapidly changing cities, control of land has also become a bone of contention. The decision to allow the Custom House Docks Development Authority to exercise planning powers in the docklands area between 1987 and 1997 undermined and sidelined the traditional role of the local authority, Dublin Corporation in this part of the city and formed part of what might be termed a broad push towards the development of the neo-liberal city. The inherent difficulties this created were widely reported and only ameliorated by the decision of central government not to transfer automatically planning powers to the DDDA when the docklands area was extended. At present, the DDDA have planning powers in a limited number of areas and a growing culture of cooperation is emerging between the docklands and local authorities in the last decade; the relationship has not evolved so easily with other major institutional landowners in the vicinity, particularly the Dublin Port Company. As had happened in 1982 when the Custom House Docks site was nationalized, the Dublin Port Company in 1997 were again perturbed when they were ordered to transfer valuable land (campshires) at what they considered was below market value to the DDDA to facilitate comprehensive redevelopment of the riverside.

But the most surprising challenge that has emerged in docklands has been how best to manage the relationship between various State bodies and agencies. Contrary to the general theories of urban redevelopment that suggest that powerful groups generally work together towards an agreed goal, the situation has been quite different in Dublin. The dispute between the DDDA and the Spencer Dock Development Consortium, in which the national public transport company (CIÉ) was a key stakeholder, highlights the intensely political and highly complex institutional nature of major urban redevelopment projects. The legacy of contamination in the Grand Canal Dock area, formerly under the control of Bord Gáis and by extension to the Department of Communications, the Marine and Natural Resources, had to be dealt with by the Dublin Docklands Development Authority, answerable to the Department of the Environment. Perhaps most surprising has been the obvious power and lack of accountability of the DDDA (which until recently was not covered by Freedom of Information legislation), demonstrated during discussions over the redevelopment of Stack A. Official correspondence would suggest that the DDDA were able to force the Government into making particular decisions to suit the timeframe of the DDDA rather than the public interest. In what would appear to be a clear case of the tail wagging the dog, the unlimited and uncontrolled power of the authority within the docklands raised questions, not just from lobby groups but even from within some Government Departments. Yet while navigation of the complex institutional landscape that has emerged in docklands has generated its own particular challenges, these are very different from those initially experienced. The difficulties encountered during the tenure of both the CHDDA and the DDDA have arisen from different visions of how redevelopment should be managed and controlled rather than from how it should be initiated, as was the case in the early 1980s.

Strongly influencing how redevelopment has emerged in docklands has been the changing role of the State and a continually evolving policy environment. While the initial funding provided by central government for the redevelopment of the Custom House Docks zone was allocated to the local authority for social projects as part of the Gregory Deal, this was a one-off decision motivated by attempts to gain power at any cost. More recent interventions have been less overtly political, but in some ways they have been more ideological. While the British and US governments embraced and publicized the creation of Urban Development Corporations as a method of 'rolling back the State', what it represented in Ireland was significantly more State intervention. The key difference was that rather than the local authority mediating the relationship between central government and the market, this middle tier was removed. In Ireland, central government intervened across a range of policy domains, through the provision of fiscal and other financial incentives as well as through the creation of a new regulatory environment to facilitate private investment and development. This pro-market intervention stands in direct contrast to the free market ideology that capital should be allowed to operate without constraints. However, it does fit with the core policies of neo-liberalism that promote limited intervention in decision-making generally, except when it comes to major infrastructural projects such as road-building or regeneration projects.

The radical physical and social changes that have occurred in Dublin docklands have been a direct outcome of government policies that encouraged private developers and facilitated capital investment in previously high-risk locations, whether the 'undesirable' Custom House Docks in the 1980s or the contaminated Grand Canal Dock more recently. Yet as changes have occurred in the broader social, economic and political environments, the State has shown an ability to respond flexibly and alter the relative balance of these public–private partnerships. The push for development of any kind as a way out of recession in the 1980s elevated the private sector to the key position within the development system, dictating the direction in which the docklands project evolved. The more recent general 'turn to community' at an international level and the strengthening of a national discourse around social partnership in Ireland has resulted in a re-balancing of the relationship between each of the influential actors within the urban system. Private capital has had to become more responsible to the public sector as a result of policy shifts over the last three to four years that may be reflective of a greater awareness of corporate social responsibility as opposed to the 'greed is good' philosophy of the 1980s. While this is possible, it may more likely reflect the relative scarcity of prime development sites remaining in the city, giving public agencies more leverage over private interests. A clear example emerged during the Spencer Dock debates, where the determination of the DDDA to rein in the excessive development proposals of the development consortium forced a compromise. While this may be an interesting example of the potential that the public sector has to intervene and direct private capital in the interests of some greater good, this is only possible during an economic boom. Within the context of a buoyant property market this kind of regulatory intervention can be undertaken without undue risk, but it will be interesting to see how this evolves when the economy slows and the property market becomes more uncertain.

Changes in the organization of planning and development within docklands have also had an impact at the broader city level. The growing entrepreneurialism of Dublin City Council since the late 1980s may have emerged from an awareness by the local authority of the need to alter their operating procedures to ensure other parts of the city were not adversely affected by the attractiveness of the docklands project. The local authority responded by mirroring the activities of the Custom House Docks Development Authority and adopting a more business-like approach in dealing with the private sector. Urban politics in the last twenty years has thus been characterized by an increasingly entrepreneurial spirit rather than by reactive regulation, not just in Dublin but on the international stage. In Dublin this new approach has been facilitated by the establishment of strategic units within the local authority such as the Economic Development and Corporate Policy Units in Dublin City Council. These have engaged in aggressive city-marketing to benefit from the mobility of international development capital. The City Council (formerly the Corporation, the name change being part of this image change) has become increasingly pro-active aided by the cultivation of a new management structure and ethos. An Inner City Development Advisory Team has been established to assist private developers with proposals for vacant sites, and the Economic Development Unit has an essential role in negotiating with developers on major planning applications to speed up the application process, a little like the fast-tracking idea implemented in docklands. This change from what was previously a rigid and bureaucratic *modus operandi* may be viewed as a response to the perceived erosion of local authority power that occurred following the establishment of the CHDDA and similar organizations in the late-1980s and 1990s. This perceived attack on democracy was vigorously opposed by one senior representative of the local authority who argued in interview that:

> We [Dublin City Council] don't want these quangos set up. It is wrong of government to set up the Custom House Docks Development Authorities, Dublin Docklands Development Authorities, the Temple Bar Properties. The Custom House Docks/International Financial Services Centre did not turn out to be what it was meant to be. We'd like to prove that we can do it better.

This growing entrepreneurial approach by local authorities has been experienced throughout Europe and in particular in both the UK and Ireland, and marketing or civic boosterism is now a critical part of their remit. Around the world, many waterfront redevelopment projects have been the spur for wider urban marketing measures to attract global investment. The 1987 Planning Scheme for the Custom House Docks followed this trend, promoting a fun, people place, a cultural attraction, an urban microcosm within 11 hectares. Of critical importance was the alteration of the old industrial wasteland perception and the reconstruction of a new post-industrial, clean, high-tech image, similar to that which has been undertaken in other cities (Short, 1996). Current plans for Spencer Dock and the Grand Canal Dock have continued this trend, promising to deliver a one-stop-shop for work, rest and play. While being a dramatic improvement on the dereliction that once plagued these sites, there is a downside, the bland sameness that is beginning to characterize many parts of the city centre, a criticism that applies to the way in which many cities are reconstructing themselves in the face of global competition.

The facilities being developed and architectural styles being adopted support the construction of this new anonymous identity, erasing one kind of image and replacing it with something more sterile. While aspiring to creating a sense of place, it is difficult to envisage how a collective memory of Dublin docklands can be forged, if long-term residents find it difficult to identify with the newly emerging landscape and new residents have no allegiance to these new districts, which are not particularly distinctive at either an international or urban scale. Unlike the buildings and streets that have been razed and redeveloped with a new morphology and nomenclature, the destruction and reconstruction of memory and mental maps is not as simple. It is clear that two very distinct perceptions and experiences of place exist within Dublin docklands.

Long-term residents remember the heyday of the docks and the decline that followed and, for them, urban renewal has destroyed any remaining linkages between their families and port or manual employment. In the first decade of redevelopment, rather than drawing on and modernizing the existing skills and resource base, local initiatives and expertise were disregarded. Increasingly, local interests became hijacked by global capital understandable because of government failure to identify a set of clear objectives, rather than aspirations, for the redevelopment of the Custom House Docks Area. Speculative construction activity was encouraged, creating instability when the property market experienced a downturn. This, coupled with a lack of clarity regarding the overall aims of the renewal strategy and the enthusiasm with which market-led regeneration policies were embraced by central government, made it impossible for the development authority to resist the lure or control of international investment capital. One of the immediate results was an increase in the gap between those that Sassen (1994) would consider the valorized and de-valorized populations. In interview Seanie Lambe argued that this became very apparent during the redevelopment of the Sheriff Street Flats site when an attempt was made to re-house the existing community outside the area to make way for private speculative development.

On the other hand, new residents view what has happened as hugely advantageous for improving their residential and lifestyle choices and now see themselves as part of a cosmopolitan, international community with lifestyles that would not be out of place in London or New York. They were wooed and catered for via the development of high-quality employment and services to suit their lifestyle, while many of the long-term residents found little in the new development with which they could engage. Due to the newfound emphasis on social need in the last decade, this polarization has become less apparent perhaps due to the power exerted by community representatives on the Council of the new development authority. Social infrastructure has been increasingly prioritized, which can perhaps be explained with reference to the opportunities provided by the economic boom.

Nonetheless, close monitoring is needed to ensure that promises are maintained. The latest planning scheme published by the Dublin Docklands Development Authority for the north docks has been criticized by some community leaders for not going far enough even though it has a clear social emphasis. They argue that, amidst all the development proposed,

> an urban park is a must for the local community. I would also like to see the social agenda pushed to the front with crèches and facilities for children. It is unworkable unless the transport infrastructure goes in first. We want a firm commitment from the Government on the Luas [light rail] before development starts. These are wonderful plans, but how are people going to get in or out?
> (Gerry Fay, in *Sunday Business Post*, 4 February 2001.)

The recently opened docklands station and the firm commitment from the Government to extend the red LUAS line to the Point Depot have gone some way towards addressing these concerns and delivering the necessary infrastructure to ensure the long-term sustainability of the redevelopment project.

The controlled but creative city

One of the biggest changes in docklands, mirrored in the Liberties, Portobello, Smithfield and other parts of Dublin, is that like modern-day shopping centres, the appearance of these areas is of an unrehearsed space where people move around freely, when in fact supervision abounds everywhere. Most obviously in the new residential complexes, urban space is becoming increasingly privatized and public space is becoming more tightly controlled in stark contrast to the 'carnival atmosphere' that was intended in the original Custom House Docks planning scheme:

> The public areas surrounding George's Dock will include an unfolding pageant of events, places, shops and eating areas. As an ensemble, the components of the area around George's Dock will provide a variety of sensory delights, with interesting merchandising, street activity and opportunities for lunching, dining and entertainment that will assure vitality both day and night.
> (Benson, 1993, p. 81.)

This again is not unique to Dublin and many geographers have argued that this kind of surveillance and control has resulted in the emergence of 'privatopias' or 'scanscapes' within cities. Control is exercised through electronic means such as CCTV monitoring and people are prevented from or coerced into behaving in particular ways within these urban spaces. This is also managed through the kinds of facilities or amenities made available and the behavioural possibilities that they offer. Albeit in a different way, the strict control exercized by the DDDA over what have been considered appropriate uses for Stack A is another example of the sanitization and control exercised within the docklands landscape. The reality is that a functional, nine-to-five environment has been developed, post-modern playfulness having been usurped by modernist functionality. The image of Dublin docklands portrayed on the international stage is of a high-grade business and residential environment, but an environment devoid of any real symbolism or sense of itself. This lack of meaning has not only manifested itself in docklands, but in other parts of the city as these standardized renewal plans take over.

While this image of a 'quality' environment is important, the resultant sterility can in fact become a major disadvantage. At present there is an emerging recognition that cities must market themselves with not only the classic attributes that attract investment, such as centrality, availability of a labour force and appropriate fiscal incentives, but other 'soft factors' that add some diversity to the landscape. Until now in some parts of docklands, the landscapes created by redevelopment could be considered more theme-park than city, suffering from a human deficit and becoming functional as opposed to lived spaces. The IFSC I and II buzz with people dashing frantically to and from offices in the morning and evening rush hours and at lunchtime, but there is little public activity on the streets outside of these times. Rather than the promotion of new city-quarters to be experienced, the processes and agents of rejuvenation in Dublin docklands have produced and promoted landscapes of observation, catering to the visual senses and defying the definition of cities as

> localities marked by intricate webs of human relationships and interchange, leading in turn to their tendency continually to engender multiplicity, flux, and unexpected events or experiences.
> (Allen, 2000, p. 12.)

This will have to change to ensure the long-term sustainability of docklands as the availability of attractive residential environments, a lively cultural scene and interesting public meeting places for business and leisure are becoming central in international locational decision-making (Florida, 2002; Scott, 2006). At the Grand Canal Docks, a new performing arts centre, designed by Daniel Libeskind, will form the centrepiece of a recently opened square designed by American landscape architect, Martha Schwartz. In July 2007, local entrepreneur Harry Crosbie announced that his Point Village Company had bought the new venue, to be named Canal Street Theatre, with a view to hosting major international ballet, opera and drama performances. This means that together with the redevelopment of the Point Depot within the Point Village district on the north side of the river, this one company will control all the major entertainment venues within docklands. Earlier this year in March 2007, the Government also confirmed that the National Conference Centre will be constructed at Spencer Dock with a frontage on to the river. Close to it within the confines of the original Custom House Docks site will be the long-awaited new National Theatre, the Abbey, moving from the northeast inner city into docklands. The Government has identified a potential site at George's Dock, and established a working group consisting of the Office of Public Works, the Department of Arts, Sports and Tourism, the National Development Finance Agency, Docklands Development Authority and the Abbey to produce a design brief. The development of this new cultural infrastructure should, twenty years after the initial development proposals for docklands, firmly plant this neighbourhood on the map of Dubliners and tourists alike as a distinct destination within the city. It may also add some much-needed vibrancy to the area after hours and broaden the appeal and perception of docklands on the international stage. The delivery of these projects will also fill a significant gap in the general perception of what constitutes a successful dockland regeneration project by providing a Dublin equivalent to the Cardiff Millennium Centre or the Sydney Opera House. While the development of these major venues is to be welcomed for both docklands and the city as a whole, it does little to promote informal cultural activity at street level within the district. The aspirations of the original architect of the Custom House Docks, Benjamin Thompson, to create a people-place with an emphasis on human interaction, 'of places intimate by day and radiant by night, where the lovely unpredictability of life would be nurtured and experienced on a daily basis' (*Boston Globe*, 19 August 2002) has yet to be achieved.

Future challenges

In the tone of a literary thriller, the evolution of regeneration in Dublin docklands since the early 1980s could be described as a composite of several separate yet interconnected plots, involving networks of private capital, national economics, environmental degradation, and shifting political culture and power. While these are individually challenging, together they will culminate in one key challenge over the coming years, to ensure that the docklands becomes an integrated part rather than a place apart from the rest of the city. In general, the core principle behind waterfront rejuvenation is to re-integrate the waterfront with the city and to rid the city of a brownfield zone in transition. While there have been many successes, increasingly the docklands and other renewal projects throughout the city have borrowed little from their wider urban context. There is nothing particularly Irish about Dublin docklands and as a result of the scale and nature of redevelopment, the severed links between the port and city rather than being repaired have been copper-fastened. A sense of alienation from the surrounding urban environment permeates the early phase of the Dublin project around the IFSC, and it will be important to ensure that this pattern is not continued in the large new urban quarters emerging. Given the scale of docklands, there may also be difficulty in fostering a sense of place and thus creating a shared identity among the very different groups who now call docklands home.

While this is an international problem, it would be a mistake to assume that what has happened in Dublin docklands has been simply the Irish manifestation of some sort of broad globalization of activity and ideas and that Dublin has simply been a passive recipient of global processes. What gives the story of Dublin docklands a unique twist is the power of personality, the key role that particular individuals have adopted in the area from local and national players like Tony Gregory, Charles Haughey, Ruairí Quinn and Harry Crosbie to well-known Irish developers and investors such as Dermot Desmond. What had its genesis as a project of the Left in the 1980s through the nationalization of the Custom House Docks and in the 1990s through a Labour Party-inspired attempt to adopt a strategic approach to the entire docklands area, has actually become one of the most potent symbols of Celtic Tiger Ireland. It has recently been reported that many of the large buildings in docklands now constitute significant investment properties for Irish individuals and institutions rather than the global investment companies that are often the major players in other cities. In April 2004, one newspaper outlined the close relationships and connections between a range of major players in docklands:

> The distinctive Citigroup headquarters on the Liffey was sold off in tranches, and is believed to have been bought by EBS, former AIB chairman Lochlann Quinn and another unnamed partner. Quinn is also believed to own the Harbourmaster 1 building opposite Connolly Station with a number of other investors. Harbourmaster 2 is jointly owned by Hibernian and building contractor Bernard McNamara … FBD [Insurance] still owns Harbourmaster 5. Chartered accountant Kevin Warren's clients are amongst the biggest property owners in the area, having bought La Touche House and AIB International House for a combined total of over €160 million. Warren's clients are also believed to own Guild House, which is the headquarters building of Commerzbank. The other chartered accountant with a significant presence in the IFSC is Derek Quinlan, whose clients are believed to own 2–6 Georges Dock. The Quinlan Partnership was also one of the developers of the Forum building near Jurys Inn … The other members of the consortium are developer David Arnold and Alanis. The A&L Goodbody building was developed by Alanis and is now owned by a number of clients, including those of Quinlan, while the Bank of Ireland Security Services building, New Century House, was developed as a partnership between Alanis and Derek Quinlan. Alanis, the Kelly family and Pierse developed the Clarion Quay apartment complex. They, along with hotelier Brendan Curtis, own the Clarion Hotel. They also built the National College of Ireland, which opened in the IFSC in 2002.
> (*Sunday Business Post*, 4 April 2004.)

While patterns of contemporary land ownership and investment are complex, they are clearer than those that existed during the initial phases of development. Given what is known about the way in which planning operated in Dublin in particular in the late 1980s, it would be no surprise if further information emerged from the Tribunals of Inquiry to shed light on the process of decision-making that informed the early development of the waterfront, particularly at the Custom House Docks. Already questions have been raised at the Moriarty Tribunal regarding payments in 1989 from one member of the Custom House Docks Development Company to the Taoiseach of the day.

Having identified many key issues and areas for future investigation in relation to Dublin, the experience of regeneration in the capital city can provide many lessons. As other Irish cities begin to embark on strategic waterfront regeneration programmes, many of the same challenges that have been encountered over the last twenty years in Dublin will emerge. In Cork, a major project to develop 162 hectares of riverside land has been announced that will promote employment and boost the residential community in the former docklands. The key aspirations of that Master Plan are very similar to those that have formed the focus of redevelopment in Dublin; yet, some of the same controversies are already beginning to emerge in relation to the development of high-rise buildings within the area The debates and inter-institutional difficulties that marred many proposals in Dublin are already rearing their head in Limerick as the 16-hectare docklands initiative develops. In the last year, legal challenges have been made to the attempts by the Shannon Foynes Port Company to dispose of some land within the area to facilitate redevelopment. The potential success of the Riverside City proposal will depend on adequate resolution of these disputes to ensure developer confidence.

But the story of regeneration in Dublin docklands is far from over and the area is perhaps entering one of its most exciting phases as development moves increasingly downstream. Major new developments at North Wall Quay, the Point Village and the Poolbeg Peninsula are either underway or due to begin very shortly and will raise challenges in relation to the suitability of apartments and high-rise buildings for family living, the optimum way of providing environmental infrastructure and the management of a relationship between the docklands as residential area and the needs of a functional port. The recent speculation about a possible move to Bremore near Drogheda, casts doubt on whether the port will even remain in Dublin City.

While these are important issues, the greater challenge will be to ensure that the legitimacy of the docklands project is guaranteed as the power of the DDDA is gradually withdrawn from the area. The long-term sustainability of the project will depend on the success of current initiatives aimed at promoting active citizenship and a sense of local ownership. It is likely that within the next decade the DDDA in its current guise will be disbanded as its official remit runs out in 2012. The real test of the long-term impacts of rejuvenation, and in particular social regeneration, will come when the management of this part of the city returns to an equal footing with the rest of the urban area. Dublin docklands today is a very different place to that which existed twenty years ago, and the turnaround physically, economically and socially has been remarkable; but evaluating whether the DDDA have achieved their aspiration of developing docklands as 'a paragon of sustainable inner city regeneration' (DDDA, 2003, p. 3) requires a much longer term perspective.

Bibliography

Allen, K. (2000) *The Celtic Tiger? The myth of social partnership*, Manchester: Manchester University Press.

Bannon, M. (1989) *Planning: the Irish experience, 1920–1988*, Dublin: Wolfhound Press.

Bartley, B. and Shine, K.T. (2002) Competitive city: governance and the changing dynamics of urban regeneration in Dublin. *In:* Swyngedouw, E. Moulaert, F. and Rodriguez, A. (eds), *Urbanizing globalisation: urban redevelopment and social polarization in the European city*, Oxford: Oxford University Press, pp 145–66.

Benson, F. (1988) The Custom House Docks. *In:* Blackwell, J. and Convery, F. (eds), *Revitalizing Dublin: what works?* Dublin: REPC, pp 90–8.

Benson, F. (1993) Public – private sector partnerships: the Custom House Docks – a case study. *In:* Bruttomesso, R. (ed.) *Waterfronts: a new frontier for cities on water*, Venice: International Centre for Cities on Water, pp 76–83.

Beaverstock, J.V., Smith R.G. and Taylor, P.J. (1999) A roster of world cities, *Cities*, 16 (6) pp 445–58.

Brady, J.E. (1988) Population change in Dublin 1981–1986, *Irish Geography*, 21 (1), pp 41–4.

Brady, J. and Simms, A. (eds) (2001) *Dublin through space and time*, Dublin: Four Courts Press.

Breen, A. and Rigby, D. (1996) *The new waterfront: a worldwide urban success story*, New York: McGraw Hill.

Brooking, C. (1728) A map of the city and suburbs of Dublin. Reprinted in Craig, M. (1983) *Charles Brooking, the city of Dublin 1728*. Dublin: Irish Architectural Archive.

CHDDA (1987) *Custom House Docks planning scheme*.

CHDDA (1990) *Annual report and accounts*.

CHDDA (1994) *Planning scheme*.

Cox, R.C. (1990) *Bindon Blood Stoney: biography of a port engineer*, Dublin, Institute of Engineers of Ireland.

Creamer, C. (1998) Public participation in the Custom House Docks. Unpublished MA thesis, Maynooth: NUI Maynooth.

Crickhowell, N. (1997) *The Opera House Lottery: Zaha Hadid and the Cardiff Bay project*, Cardiff: University of Wales Press.

Daly, M. (1984) *Dublin: the deposed capital: a social and economic history, 1860–1914*, Cork: Cork University Press.

DDDA (1997) *Dublin Docklands Area master plan*.

DDDA (1999) *Annual report and accounts 1998*.

DDDA (2002) *Annual report and accounts 2001*.

DDDA (2003) *Dublin Docklands Area master plan*.

DDDA (2005) *Master Plan monitoring report*.

DDDA (2007) *Annual report and accounts 2006*.

De Courcy, J. (2000) Bluffs, bays and pools in the medieval Liffey at Dublin, *Irish Geography*, 33 (2), pp 117–33.

Department of the Environment (1996) *Study on the Urban Renewal Schemes (prepared by KPMG in association with Murray O'Laoire Architects and Northern Ireland Research Centre)*, Dublin: Stationery Office.

Dublin Corporation (1986) *The Inner City draft review*, Dublin: Dublin Corporation.

Dublin Port and Docks Board (1980) *The Custom House Docks Development Proposal.*

EU Expert Group on the Urban Environment (2001) *Towards more sustainable urban land use*, Brussels: European Commission.

Event Ireland (2002) *Feasibility study for a museum of Dublin in Stack A.*

Florida, R. (2002) *The rise of the creative class and how it's transforming work, leisure, community and everyday life*, New York: Basic Books.

Gilligan, H.A. (1988) *A history of the port of Dublin*, Dublin: Gill and Macmillan.

Guy, C. (1994) *The retail development process*, London: Routledge.

Haider, D. (1992) Place wars: new realities of the 1990s, *Economic Development Quarterly*, 6 (2), pp 127–34.

Hamnett, C. (1984) Gentrification and residential location theory: a review and assessment. *In:* Herbert, D.T. and Johnston, R.J. (eds), *Geography and the urban environment: progress in research and applications, Volume 6*, London: Wiley, pp 283–319.

Harvey, D. (1989) From managerialism to entrepreneurialism: the transformation in urban governance in late capitalism, *Geografiska Annaler*, 71B (1), pp 3–17.

Hobson, B. (1930) *A book of Dublin*, Dublin: Dublin Corporation.

Hogan, J. (2006) The politics of urban regeneration, *Progress in Irish Urban Studies*, 2, pp 27–37.

Hoyle, B.S. and Hilling, D. (1984) *Seaport systems and spatial change: technology, industry and development strategies*, London: Wiley.

Hoyle, B.S. *et al.* (1994) *Revitalizing the waterfront: international dimensions of dockland redevelopment*, London: Belhaven Press.

IDA Ireland (1997) *Achieve global competitive advantage in financial services.*

Kohnstamm, P. (1993) Urban renewal and public-private partnership in the Netherlands. *In:* Berry, J. *et al.* (eds) *Urban regeneration: property investment and development*, London: E&FN Spon, pp 220–9.

Kyne, D. (1989) An evaluative study of three dockland redevelopments: results and possible applicability to Dublin. Unpublished MRUP thesis, Dublin: UCD.

Logan, J.R. and Molotch, H.L. (1987) *Urban fortunes: the political economy of place*, Berkeley, CA: University of California Press.

MacLaran, A. (1993) *Dublin: the shaping of a capital*, London: Wiley.

Malone, P. (1996) *City, capital and water*, London: Routledge.

Markuse, P. and vanKempen, R. (2000) *Globalizing cities: a new spatial order?* London: Blackwell.

Maxwell, C.E. (1997) *Dublin under the Georges, 1714–1830*, Dublin: Lambay Books.

McCarthy, M. (ed.) (2005) *Ireland's heritages: critical perspectives on memory and identity*, Aldershot: Ashgate.

McDonald, F. (2000) *The construction of Dublin*, Oysterhaven, Kinsale: Gandon Editions.

McGuirk, P. (1995) Power and influence in urban planning: community and property interests' participation in Dublin's planning system, *Irish Geography*, 28 (1), pp 64–75.

McIntyre, O. (2003) Problems of liability for historical land contamination under Irish law, *Irish Planning and Environmental Law*, 10 (4), pp 112–18.

McManus, R. (2002) *Dublin, 1910–1940: shaping the city and suburbs*, Dublin: Four Courts Press.

McManus, R. (2005) Identity crisis? Heritage construction, tourism and place marketing in Ireland. In: McCarthy, M. (ed.) *Ireland's heritages: critical perspectives on memory and identity*, Aldershot: Ashgate, pp 235–50.

Moore, N. and Scott, M. (2005) *Renewing Urban Communities*, Aldershot: Ashgate.

Moore, N. (1999) Rejuvenating docklands: the Irish context, *Irish Geography*, 32 (2), pp 135–49.

Murphy, D. (2002) *Ireland and the Crimean war*, Dublin: Four Courts Press.

Newman, O. (1972) *Defensible space: crime prevention through urban design*, New York: Macmillan Company.

North Inner City Folklore Project (1995) *Reminiscences north of the Liffey*, Dublin: North Inner City Folklore Project.

O'Donovan, J. (1986) *Life by the Liffey: a kaleidoscope of Dubliners*, Dublin: Gill and Macmillan.

O'Keeffe, T. (2005) Heritage, rhetoric, identity: critical reflections of the Carrickmines Castle controversy. In: McCarthy, M. (ed.) *Ireland's heritages: critical perspectives on memory and identity*, Aldershot: Ashgate, pp 139–51.

Oram, H. (1998) Docklands revival, *Technology Ireland*, pp 30–3.

Pahl, R. (1975) *Whose city?: and further essays on urban society*, London: Penguin.

Prunty, J. (1988) *Dublin slums, 1800–1925: a study in urban geography*, Dublin: Irish Academic Press.

Prunty, J. (1995) Residential Urban Renewal Schemes, Dublin 1986–1994, *Irish Geography*, 28 (2), pp 131–49.

Rybczynski, W. (2006) Shipping news, *New York Review*, 10 August, pp 22–5.

Sassen, S. (1994) *Cities in a world economy*, California: Pine Forge Press.

School of Architecture UCD (1996) *Inventory of the Architectural and Industrial Archaeological Heritage, 2 volumes*, Dublin: CHDDA.

Scott, A.J. (2001) Capitalism, cities and the production of symbolic forms, *Transactions of the Institute of British Geographers*, 26 (1), pp 11–23.

Scott, A.J. (2006) Creative cities: conceptual issues and policy questions, *Journal of Urban Affairs*, 28 (1), pp 1–17.

Short, J.R. (1996) *The urban order: an introduction to cities, culture and power*, Cambridge: Blackwell.

St Andrews Heritage Group (1992) *Along the quays and cobblestones: folklore of the south docks community*, Dublin: St Andrews Heritage Group.

Telesis Consultancy Group (1982) *A review of industrial policy: a report prepared by the Telesis Consultancy Group*, Dublin: National Economic and Social Council Reports, no. 64.

Wright, G.N. (1821) *An historical guide to ancient and modern Dublin: illustrated by engravings after drawings by George Petrie, esq, to which is annexed a plan of the city*, London.

Illustrations

1	The Liffey channel in the seventeenth century.	17
2	The bay and harbour of Dublin.	18
3	The development of Dublin's quays by 1728.	19
4	Custom House.	20
5	Dublin port from Essex Bridge to the Barr, 1704.	21
6	Engineering works underway on the channel, 1756.	22
7	The 'lotts' and the docklands in 1836.	23
8	Newfoundland Street, 1876.	24
9	Dublin port improvements, Captain Bligh, 1803.	25
10	Chart of Dublin Bay with cross-sections of the wall, 1881.	26
11	Dublin port in 1846.	27
12	Shipping near the Custom House.	28
13	Port and harbour of Dublin, 1887.	30
14	Industrial uses in south Docklands, 1911.	32
15	Clayton gasometer, Barrow Street, prior to redevelopment.	33
16	Port activity in early 1920s.	34
17	Advertisement for the port of Dublin.	35
18	Aerial view of Dublin docklands, 2004.	36
19	Busárus, Beresford Place, 1960s.	36
20	Old Custom House Dock.	37
21	The extent of the Dublin Docklands Area.	38
22	Saltaire near Bradford.	41
23	Abandoned warehouses in Liverpool, 1980s.	43
24	The extent of reclaimed land in Dublin Port.	45
25	Manual cargo handling at the B&I terminal, North Wall Quay.	45
26	Container handing in Dublin Port, 2003.	46
27	Docklands at Melbourne in 1997 as renovation begins.	47
28	A view down the Liffey in the early twentieth century.	48
29	A view down the Liffey in 1976.	48
30	Dereliction, Summerhill, 1983.	50
31	Dereliction, Mountjoy Square, 1983.	50
32	Aerial photograph of Sheriff Street flats, 1994.	51
33	Local authority housing at Thorncastle Street, Ringsend.	52

ILLUSTRATIONS

34	Industrial and residential landuses in close proximity.	53
35	Housing quality in Dublin Docklands, 1925.	54
36	Outline diagram of redevelopment, Sydney.	59
37	Darling Harbour, Sydney.	59
38	Darling Harbour after redevelopment in 1998.	60
39	Bristol Harbour, the Arnolfini centre for contemporary arts.	62
40	South Street Seaport, New York.	62
41	Gateways and hubs in the National Spatial Strategy.	69
42	An Taisce publication opposing the port proposals, 1973.	72
43	The Irish Life Centre, Abbey Street, under construction, 1977.	74
44	The Docklands delimited by the construction of the Talbot Memorial bridge, 1980.	74
45	Structure of the Custom House Docks site.	75
46	New office block on Mespil Road, 1982.	77
47	Need for renewal on St Stephen's Green, 1983.	77
48	Parnell Street, site of the former Williams and Woods factory, mid 1980s.	78
49	Media reaction to the 'Gregory Deal'.	81
50	Local authority housing development at City Quay.	86
51	Local authority housing on Townsend Street.	86
52	Cambridge Court, Ringsend.	87
53	Opposition to the Urban Development Areas Bill.	87
54	Cumberland Street, 1980s.	90
55	Dublin quays, 1987.	90
56	Map of urban renewal area.	95
57	Advertisement for urban development site, inner city Dublin, 1990.	95
58	USS *Constellation* and the inner harbor Baltimore.	98
59	Canary Wharf development, London.	98
60	Techniquest science museum, Cardiff.	103
61	Artist's impression of docklands redevelopment.	103
62	This office block on Mountjoy Square seemed destined never to be completed, 1990.	105
63	Aerial view of docklands, 1986.	106
64	Derelict interior of Stack A warehouse.	107
65	Process of extension of Custom House Docks area.	112
66	Demolition of stacks for the IFSC development, 1988.	115
67	Block 1 nears completion, 1989.	115

68	The Liffey vista by 1990.	117
69	Nearby development – Irish Life, Abbey Street, 1990.	117
70	The George's Quay site, adjacent to the IFSC, had lain undeveloped for most of the 1980s.	119
71	Dublin Exchange Building, Custom House Docks.	120
72	ABN Amro Building.	122
73	Aerial view of docklands, 1994.	125
74	Aerial view of docklands, 1997.	127
75	Completed projects within the Custom House Docks, 1997.	128
76	The Harbourmaster Bar, Custom House Docks.	130
77	Jury's Custom House Inn, Custom House Quay.	130
78	Fisherman's Wharf, Ringsend.	132
79	Custom House Docks apartment development.	134
80	Security at Custom House Docks.	134
81	Juxtaposition of Custom House Docks development and Sheriff Street flats, 1996.	136
82	Sherriff Street, 1997.	137
83	New public housing at Sheriff Street.	139
84	Outline diagram of redevelopment showing public park.	141
85	Playground as a buffer zone, Sheriff Street area, aerial view, 2005.	142
86	Playground as a buffer zone, Sheriff Street area, 2005.	142
87	Custom House Square, Mayor Street.	144
88	Mayor Square and environs.	144
89	The 'wall' between Sheriff Street and the IFSC, 1999.	146
90	The 'wall' between Sheriff Street and the IFSC, 2005.	146
91	Scale of the extended docklands area from 1996.	151
92	New DART station at Barrow Street.	153
93	East Point Business Park, aerial view, 2004.	156
94	The Wiggins Teape factory, East Wall, prior to demolition, 2005.	156
95	The pre-development landscape around George's Quay.	157
96	George's Quay in 1990.	157
97	The first phase of development nears completion on George's Quay, 1992.	158
98	The completed development on George's Quay – the 'canary dwarfs'.	158
99	Pearse Square	161
100	Gilbert Library and environs, Pearse Street.	161
101	Docklands district electoral divisions.	167

102	Changing age structure in Dublin docklands.	168
103	Café culture in Dublin docklands.	169
104	New commercial outlets on Mayor Street.	170
105	Changing unemployment rates in Dublin docklands.	171
106	Apartment sales leaflet, 1998.	175
107	Changing levels of highest educational achievement within docklands.	185
108	National College of Ireland.	190
109	Dock Mill apartments, 2005.	194
110	Clarion Quay, completed development, 2005.	197
111	First phase of Gallery Quay nears completion, 2004.	197
112	Hanover Quay, late-2004 – Grand Canal Basin.	203
113	Forbes Quay under construction, November 2004.	203
114	Hanover Quay, dockside, with Longboat Quay, October 2005.	204
115	Longboat Quay apartments, July 2007.	204
116	Clarion Quay apartments.	206
117	George's Dock and the Stacks in use, 1977.	217
118	Stack A in 1999.	217
119	Artist's impression of Stack A redevelopment.	222
120	Renovation commences 2000.	222
121	Conservation work on Stack A.	223
122	*chq* brand applied to Stack A, 2005.	232
123	Renovated Stack A, 2005.	232
124	Stack A and George's Dock, 2005.	233
125	Spencer Dock environs, 1994.	238
126	Spencer Dock environs, 2005.	238
127	Spencer Dock environs, 2005.	239
128	Proposed locations for the National Conference centre.	241
129	Advertising banner for Spencer Dock development.	244
130	Schematic drawing of the original Spencer Dock proposal.	244
131	The Spencer Dock development in its context.	245
132	Older residential development, Upper Mayor Street and environs.	245
133	The development process at Spencer Dock.	253
134	The new Liffey vista, north quays, 2005.	255
135	Northwestern Railway Hotel, early-twentieth century.	257
136	Northwestern Railway Hotel in 2005.	257
137	Revised Spencer Dock scheme.	258

138	Revised Spencer Dock scheme and old Northwestern Railway Hotel, 2005.	259
139	Aerial view of the Spencer Dock scheme, 2005.	260
140	Selling the Spencer Dock scheme, 2006.	261
141	Selling the Spencer Dock scheme, 2006.	261
142	Spencer Dock scheme, summer 2007.	262
143	Gas production, Grand Canal Docks, 1983.	264
144	Grand Canal Docks derelict site, 1994.	264
145	Grand Canal Docks derelict site, 1994.	266
146	Grand Canal Docks derelict site, 1994.	266
147	Original Grand Canal Docks development scheme.	269
148	Aerial view of Grand Canal environs, 1997.	270
149	Rapid pace of development, 2005.	271
150	Aerial view of development, January 2005.	272
151	View from the south, May 2005.	273
152	Pearse Square, Pearse Street, overshadowed by new development, 2005.	274
153	The old chimney amidst new development, 2005.	275
154	Millennium Tower: High-rise apartment living, 2006.	276
155	Inner Grand Canal Quay prior to its redevelopment in 2006.	276
156	Grand Canal Square takes shape, summer 2007.	277
157	The new streetscapes of docklands, 2007.	277
158	New commercial development at South Bank Quay.	278
159	Aerial view of the Poolbeg Peninsula.	279
160	The industrial landscape of the Poolbeg Peninsula.	279
161	Local authority housing on the south-western edge of the Poolbeg Peninsula on reclaimed land.	280

Index

Numbers in italics refer to illustrations

Abbey Theatre, 227, 295
ABN Amro, 121
Abrams, Charles, 67–8
Act of Union (1801), 29, 40
Admiralty, 25
Affordable Housing Programme, 164
Ahern, Bertie, 138, 225, 287
AIB Bank, 113, 116, 297
AIB International House, 297
air transport, 44
A&L Goodbody building, 297
Alanis, 205, 297
Albert Docks, Liverpool, 234–5
Alexandra Basin, 31, 32, 47
Alexandra Quay, 133
Alexandra Road, 99
Alliance for Work Forum, 110, 178
Amiens Street, 97
Andrews, Chris, 283
Andrews, Niall, 89, 91
Anglo-Irish Bank, 188, 189
Anglo-Normans, 16
Anna Livia consortium, 240
AOL, 125
apprenticeships, 177
Aqua Terre Solutions, 268
archaeological interest, zones of, 160
Area Framework Plan, 155
Arnold, David, 297
Arts, Heritage, Gaeltacht and the Islands, Department of, 225–6, 229
Arts, Sport and Tourism, Department of, 228, 229–31, 295
Astons 'Key', 20

Bacon, Peter, 225
Ballsbridge, 23, 79, 198, 240
Ballymun, 208
Ballymun Regeneration Project, 259
Baltimore, 56, 58, 107, 173
    'festival marketplace,' 105
    Inner Harbor, 60, 63, *98*, *99*, 102
Bank of Ireland, 116, 194, 297
Bannon, Michael, 99–100
Barcelona, 61, 286

Barrow Street, 32, 131, 196, 198
    DART station, 153
Battery Park City, New York, 212
Belfast, 61, 227
Benjamin Thompson and Associates, 104
Benson, Frank, 220
Beresford Place, 104, 120
B&I Terminal, North Wall Quay, *45*
Birmingham, George, 88–9
Bligh, Captain John, 25, 28
Blind 'Key', 17
Book of Dublin, A, 33
Bord Fáilte, 225, 240, 246
Bord Gáis, 180, 288
Bord Pleanála, An, 79, 89, 283
Boston, 37, 105, 211
Bounty, HMS, 25
Breen, Gerry, 236
Bremore, Co. Louth, 38, 298
Bridge Street, 160
Bristol Harbour, *62*
Bristol Science Centre, 227
Britain, 41, 64, 124, 148, 163
    economic decline, 42–3
    planning, 67, 68, 70, 100, 291
    science centres, 220
    Urban Development Corporations, 93, 289
Brooks Thomas, 131
brownfield sites, 263, 265, 296
    Grand Canal Docks, 267–75
    Poolbeg Peninsula, 275–84
Buckley, Michael, 113, 116
Burke, Ray, 89
Busáras, 34, *36*, 73, 75, 76

Cambridge Court, 85, *87*
Camden, earl of, 24
Canal Street Theatre, 295
'canary dwarfs,' *158*
Canary Wharf, London, *98*, 99, 112, 116, 237, 243
Cape Town, 37, 63
capitalism, 43–4
Cardiff, 102, *103*, 151, 219–20, 227, 253
    Opera House, 237, 239–40

309

Cardiff Bay Development Corporation, 237, 239–40
Cardiff Millennium Centre, 295
cargo handling, 44, 46
Carlow Chamber of Commerce, 70
Carlton Group, 240
Carr Communications, 183
Carrickmines Castle, 214
Cashman, Donal, 85
CCTV, 175, 294
Celtic Tiger, 121, 265, 286–7, 296
   and docklands development, 37, 39, 108, 122–5
   'dual city,' 162, 174
   sources of, 148–9
Census of Population, 55, 167
Central Bank of Ireland, 124, 193
Central Business District (CBD), 73, 75
Centre for Childcare and Adolescent Studies, 191
Centre for Educational Opportunity, 191
Channel Islands, 114
Chase Manhattan, 116
chemical industry, 32
Chesterbridge Ltd, 141, 143, 199
Childcare Forum, 208
Children's Museum, 228
chq (Stack A), 231–5
Christ Church, 16
Church Road, 159
CIÉ, 34, 121, 240–1, 274, 281
   NCC proposal, 240, 241–2
   Spencer Dock consortium, 243, 247–9, 258, 259, 262–3, 288
Cisco Systems, 125
Citibank, 116, 118, 123, 189
Citicorp, 116
Citigroup Europe, 123
Citigroup Fund, 189
Citigroup Ireland, 118, 123
City Arts Centre, 230
City Housing Initiative, 209
'City of Man,' 104, 152
City Quay, 55, 151, 154, 159, 168
   cruise liner terminal, 160
   housing, 85, *86*
City West, 126
Civic Offices, 16, 94, 215
Civic Survey, 1925, 55
Clarion Hotel, 189, 297
Clarion Quay, *197*, 205, 206–7, 210, 297
Clayton gasometer, 32, *33*
Clinton, Bill, 148, 163

Clohessy, Lewis, 223
Clúid Housing Association, 200, 205–6, 207–8
Clydeside Regeneration Project, 220
Cockle Bay Wharf, Sydney, 61
Coin Street, London, 211
College Green, 40, 271
Collins, Greenville, 16
Columbia University, New York, 68
Colvin, Liam, 194
Combat Poverty Committee, 84
Combined Cycle Gas Turbine, Ringsend, 278
Combined Residents Association, 281
Commerzbank, 123, 297
Commons Street, 97, 145, 174
communal space, 208
communication technologies, 42, 44
Communications, the Marine and Natural Resources, Department of, 288
Community Liaison Committee (CLC), 101, 150, 178, 180–4, 210–11
compulsory purchase, 84
Conference of Religious in Ireland (CORI), 174
Connolly Station, 76, 111, 121, 297
Constellation, USS, 98
containerization, 44, 46, *46*
Coopers & Lybrand, 147
Cork, 70, 85, 298
Corporate Policy Unit, DCC, 291
corporation tax, 114
Corr, Frances, 281
Coyne, Peter, 225–7, 229, 247, 268–9
Creighton Street, 85
Crickhowell, Lord, 239, 240
Crimean War, 216
Crosbie, Harry, 21, 295, 296
Crosbie's Yard, 196
cruise liners, 160
Cullen, Martin, 206
cultural development, 106–7, 147, 213–14, 284
   Grand Canal Square, 273–4
   privatization, 293–5
   Stack A, 216–36
Cumberland Street, *90*
Cunningham, Bill, 147
Curtin, Dónall, 182, 183, 184, 185, 193, 211
Curtis, Brendan, 297
Custom House, 23, *28,* 29, 34, 76, 104–5
   new location, 20
Custom House Docks, 29, 31, 34, *37,* 149, 151, 152, 154, 188; *see also* Stack A warehouse
   1987-97, 109–62
   and Celtic Tiger, 122–5
   completed projects, *128*

development plans, 71, 73, 75–80, 266
education, 185
funding, 289
Gregory Deal, 80–92
lack of amenities, 129, 147
and local housing, 135–48
nationalization, 83, 89, 96, 135, 254, 287, 288, 296
objectives, 101–8, 292, 293
other elements of scheme, 129–31
plan of site, *75*
planning environment, 97–101, 155
relocations from, 271, 273
tax incentives, 93–4
Custom House Docks Development Authority (CHDDA), 94, 96–7, 109, 116, 178, 211, 289
area covered by, 97, 99, 121
commercial development, 122, 129
and Dublin Corporation, 288, 291
employment, 171–2
extension plans, *112*
failures of, 149–50, 162, 179
housing, 137, 140–8
and IFSC, 120
and local community, 176–9, 181, 182, 287
marketing, 172
Master Plan, 104–8, 111, 127, 145, 216, 218
planning powers, 236
and private sector, 290
profits, 124
role of, 99–101
Stack A, 216, 218, 220–5, 231
tax incentives, 126
and tribunals, 297
Custom House Harbour, 133, 135, 145, 174
Custom House Plaza, 121, 174
Custom House Quay, 17, 97, 104
Custom House Square, 143, *144*, 199

Dáil Éireann, 84, 99, 122
Urban Development Areas Bill, 88–91
Urban Renewal Act, 92, 94, 96–7
Dame Street, 40
Daninger, 131
Darling Harbour, Sydney, *59, 60*
DART station, Barrow St, 153, 198
DART station, Grand Canal Dock, 274, 293
de Rossa, Proinsias, 96–7
Deasy, John, 66
decentralization, 70, 79
Deloitte & Touche, 220, 221
Democratic Left, 163

Dempsey, Noel, 182, 228
derelict site tax, 81
Derelict Sites Act, 1991, 265
Desmond, Barry, 114
Desmond, Dermot, 39, 113, 296
and IFSC, 110–11, 116, 118, 124, 147
and Spencer Dock, 250, 251–2
and Stack A, 224–5
Discovering University programme, 191
Dock Mill apartments, *194*, 196, 198
Dockland Area Bus System (DABS), 153–4
docklands social regeneration conference, 183–4
Dodder, river, 25, 154
Donnelly, Professor Dervilla, 224
Donnybrook Bus Garage, 73
Dowling Court, 85
Drogheda, Co. Louth, 38
'dual cities,' 64
Dublin, 70, 89, 286
eastern movement, 73, 75
historic core, 94
need for regeneration, 66–7, 79–80
population, 49
quays, *19, 90*
Dublin: A City in Crisis (RIAI), 92–3
Dublin Bay, 16, *18*, 21, 24, 281
Dublin Bus, 248
Dublin Chamber of Commerce, 180
Dublin Chronicle, 24
Dublin City Council; *see* Dublin Corporation/City Council
Dublin City Council Shared Ownership Scheme, 208–9
Dublin City Development Plan, 2005-2011, 282
Dublin City University (DCU), 220
Dublin Corporation/City Council, 34, 38, 68, 80, 89, 153, 154, 159, 181
Area Framework Plan, 155
and CHDDA, 100–1, 288
Development Plan, 100
Dublin Port and Docks Board proposals, 73, 75–80
Dublin Museum, 228–9
entrepreneurial, 290–1
and historic core, 94
housing, 81–2, 83, 85, 202, 211
docklands, 137–41, 143
name change, 290
and planning, 236–7
and Poolbeg Peninsula, 282–4
security, 174
and Spencer Dock, 246, 249–50, 255
Dublin County Borough, 55

Dublin Crisis Conference, 93
Dublin Docklands, 39–40, 160, 162
  1836, *23*
  aerial views, *36, 106, 127*
  continuing evolution, 286–99
  development, 154–60, 296–9
    1997 plan, 151, 152–4
    challenges, 287–93
    changing directions, 163–212
    and 'community capacity,' 176–84
    loss of identity, 291–2
    model of good practice, 211–12
    obstacles, 213–85
    politics of, 236–42, 252–6
    politics of planning, 66–108
    private sector interest, 109–11
    problem sites, 263–7
    questions, 284–5
    role of IFSC, 111–25
    role of State, 289–90
    Urban Renewal Act, 92–7
  'dual city,' 162, 174
  education, 184–95
  historic context, 15–16
  housing, *54*
    apartment developments, 195–211
  industry, *53*
  land ownership, 288
  local population, 109–10, 292–3
    social change, 165–72
  marginalization, 49
  marketing, 172–6
  official extent of, 37
Dublin Docklands Development Authority Act, 180, 202
Dublin Docklands Development Authority (DDDA), 38, 123, 154, 295
  and CIÉ land, 242
  Council, 179–81
  disbandment, 298–9
  education, 184–95
  employment, 171
  established, 148–51
  future plans, 154–60
  lack of planning powers, 236–7
  and local authority, 288, 291
  and local community, 177–84, 287
  marketing, 172–6
  Master Plan, 151, 152–4, 164, 179, 186–7, 188, 191, 193, 256, 281
    2003, 154, 155, 196
    environment, 263, 265
    housing, 196, 202, 205, 208, 209

incinerator, 282
  Spencer Dock, 246–8
  and NCI, 187–95
  and private developments, 281, 284, 285
  problem sites, 263–7
  and Spencer Dock, 243–56, 254, 255
    Section 25 scheme, 256, 259
  and Stack A, 218, 225–31, 288–9, 294
Dublin Exchange Facility, 120–1
Dublin Gas Act (1866), 32
Dublin Harbour, map, *21*
Dublin Interactive Science Centre (DISCovery) group, 220–8
Dublin Metropolitan Streets Commission, 101
Dublin Port, *27*, 28–9, 44, 280
  1887, *30*
  advertisement for, *35*
  chart, *26*
  containerization, 46
  decline, 1980s, 49–56
  deep-water berthage, 31, 33–4, 47
  industries in, 47, 49
  migration eastward, 21–5
  move planned, 17, 20, 38, 298
  navigation channel, 20–1
  need for regeneration, 79–80
  reclaimed land, *45*
  1920s, 32–3, *34*
  shipwrecks, 21
  trade, 29, 31
Dublin Port and Docks Board, 31, 33
  and campshires, 179
  and nationalization, 83, 84–5, 135, 254
  redevelopment plans, 71–9, 81, 287
  right to sell lands, 78
Dublin Port Company, 20–1, 38, 288
Dublin Service Centre, 123
Dublin Tourism, 221, 223, 224
Dublin Transport Authority, 101
Dublin Walled City Development Authority, 94
Dublin Waste Management Plan, 282
Dublin West constituency, 94, 96
Duffy Lally development, 274
Duggan, Fr Frank, 109–10
Dun Laoghaire, Dublin, 63
Dun Laoghaire-Rathdown County Council, 68, 283
Dundrum Town Centre, 273
Durney, Terry, 221
Dwyer, Dermot, 247

East Link Bridge, 131–2, 278
East Point Business Park, 125, 126, 155, *156*

East Point Enterprise Zone, 125
East Road, 209
East Wall, 21, 52, 151, 154, 155, *156*, 159, 184, 209, 249
    education, 192
    and Spencer Dock, 262
East Wall Community Development Council, 176, 183
East Wall Enterprise Zone, 126
East Wall Residents' Association, 178
East Wall Road, 31
Economic and Social Research Institute (ESRI), 165, 185–6
Economic Development Unit, Dublin City Council, 290–1
education, 184–95
Education, Department of, 187–8
Educational Building Society (EBS), 297
Eircom, 125
Electricity Supply Board (ESB), 278, 280
Ellier Developments, 274
Ellis, William, 16
Ellis Quay, 17
Ely Restaurant and Wine Bar, 233
emigration, 124
employment, 152–3, 171–2, 177, 178
English Heritage Trust, 234
Enterprise, Trade and Employment, Department of, 227
Enterprise Zones, 125
Environment, Department of the, 94, 135, 149, 180, 199, 206, 211, 288
    apartment guidelines, 200–1
    and DDDA, 229
    Green minister, 214
    Sheriff St, 137–9
    Spencer Dock, 247
environmental problems, 263–7
    Grand Canal Dock, 267–75, 288
    Poolbeg Peninsula, 275–84
Environmental Protection Agency, 269, 270
EOLAS, 220
Essex Bridge, 40
EU Expert Group on the Urban Environment, 263–7
Europe, 42–3, 116, 131
European Commission, 240, 243, 248, 283
European Union, 114, 240, 254
    and Celtic Tiger, 149
    new social agenda, 163
    and tax incentives, 126
Event Ireland, 235

Fabrizia, 131
FÁS, 110, 194
Fay, Gerry, 138, 143, 181, 184, 210, 250, 293
FBD Insurance, 297
Feely, Frank, 78, 137–8
Ferryman hotel, 271
festival marketplaces, 61, 63, 102, 105
Fianna Fáil, 82, 83, 91, 96, 113, 283
Finance, Department of, 113
Finance Act, 1987, 93
Finance Dublin Yearbook 2006, 124
Fine Gael, 81–2, 88, 94, 113, 163
FINEX Europe, 120–1
Fingal County Council, 68, 283
First National Building Society, 118
Fisherman's Wharf, 132–3
FitzGerald, Alexis, 89
FitzGerald, Garret, 80, 81–2, 93
Fitzpatrick, Sean, 188
Flanagan, O.J., 82–3, 88, 122
flooding, 24
Flynn, Chris, 229
Flynn, Pádraig, 96, 97, 138
Forbes magazine, 110
Forbes Quay, 202, *203*
Forfás, 121
Fortis Bank, 259
Forum building, 297
Four Courts, 16
Francis Street, 166
Freedom of Information legislation, 288
funds industry, 124

Gallery Quay, 196, *197,* 198
Gandon, James, 104
Garda Síochána, 96
gasometers, 31–2, *33,* 54, 180, *264,* 265, 266
Gasworks site, Barrow St, 131, 196
gated communities, 174, 201–2
'gateways,' 70
General Electric, 116
George IV, King of England, 28
George's Dock, 28, 29, 76, 107, 121, 227, 293, 297
    Abbey site, 295
    glazed pyramid plan, 224–5
George's Quay, 75, 97, *119, 157, 158,* 159, 160
Germany, 124, 126, 152, 286
Gilbert Library, 159, 160, *161*
Giles, Francis, 28
Glasgow, 40, 211
Glasgow Science Centre, 220
Gleesons Field, 170

'global cities,' 42
globalization, 42–4, 56–7, 63, 296
   and urban waterfronts, 44–9
Globalization and World Cities research centre, 286
Gormley, John, 214, 283
Grand Canal, 37
Grand Canal Basin, 154, *203*
Grand Canal Dock, 22, 24–5, 54, 151, 153, 154, 177, 240, 282, 285
   cultural development, 295
   DART station, 274
   dereliction, *264, 265, 266*
   difficulties, 213, 214
   environment, 288
   housing, 196, 198
   plan of development, *269*
   Planning Scheme, 155, 205, 209
   remediation and redevelopment, 267–75
Grand Canal Enterprise Zone, 126
Grand Canal Harbour, 269, 274
Grand Canal Square, 273–4, *277*
Graphic, 29
Great South Wall, 28
Green Party, 214, 242, 283
Gregory, Tony, 39, 73, 110, 139, 140, 176, 296
   DDDA Council, 181
   and education, 186, 192–3
   Gregory Deal, 80–92, 133, 254, 287, 289
   NCC deal, 241–2
   Spencer Dock, 210, 252
Guilbaud, Patrick, 199
Guild House, 297
Guild Street, 97, 250
'Gulliver,' 221, 224

Hadid, Zaha, 237, 240
Hamilton Osborne King, 174
Hanover Quay, *203, 204,* 274
Haran, Paul, 227
Harbourmaster Place, 119, 121, 297
Harbourmaster Pub, 129, *130*
Harbourside Shopping Centre, Baltimore, 63
Harcourt Street, 126
Hardwicke/British Land consortium, 119–20, 121, 218, 221, 226
Hardwicke Ltd, 124
Harvey Nichols, 233
Haughey, Charles, 39, 73, 287, 296
   Gregory Deal, 80–92, 110
   and IFSC, 112–13, 118, 124
Haughey, Seán, 101
Henrietta Street, 93–4

heritage
   and development, 214–15
   and marketing, 234–6
Heritage Properties, 273
Heuston Gate, 228
Hogan, Maurice, 113
Holiday Inn, Pearse St, 159
Hollywood, Gráinne, 226
Holyhead, 29
Hooke and MacDonald, 196, 201
housing, 20, 49, *52,* 52–3, 55–6, 80, 91
   apartment developments, 195–211
      guidelines, 200–1
   CHDDA, 105
   corporate investment, 198–9
   Gregory Deal, 81–92
   inner-city living, 131–5
   management, 211
   Poolbeg Peninsula, *280*
   quality, *54*
   social housing, 152, 179, 202, 205–8, 212
   social polarization, 173–6, 206–8
   'socio-economic cleansing,' 135–48
   Spencer Dock, 209–11, 248
   warehouse conversions, 58
Housing Forum, 211
Howlin, Brendan, 149, 150, 163, 180
'hubs,' 70
Humphries, Kevin, 176
Hussey, Gemma, 215, 235
Hyatt Residence Hotel, 273

IBEC, 180
IBI Group, 268
ICS Building Society, 194
IFSC House, 116, 118
Illustrated London News, 29
image change, 57–8
Inchicore Chassis Works, 73
incinerators, 214, 281
   Poolbeg site, 281, 282–4
industrial development, 31–2, 33–4, 36–7
   cities, 40–1
   urban waterfronts, 44–9
Industrial Development Authority (IDA), 113, 114, 116, 119, 193
Infirmary Road, 240
Inner City Development Advisory Team, 290–1
inner-city living, 131–5
'Inner City Problem, The,' 73
Inner City Trust, 178
Inner Dock, 28, 29, 105, 133, 135, 143, 145, 196
Inner Grand Canal Quay, *276*

Integrated Area Plans, 148
Interactive Science Centre, 218–28
International Financial Services Centre (IFSC), 28, *117*, 149, 213, 236, 250
   apartments, 143, 145, 199–200
   construction, *115*
   employment, 177, 178
   failures of, 251–2, 291, 294, 296
   flagship role, 101, 104–5, 111–12, 286–7
   genesis of, 112–16
   local population, 162, 168, 270–1
      'Berlin wall,' *146*
   museum plan, 223–5, 231
   NCI site, 188, 248
   ownership, 297
   politics of development, 237
   post-1990, 116–22
   rents, 118
   second phase, 122–4
   services, 129, 131
   tax incentives, 126–8
   training, 193
International Financial Services Institute, 193–4
internet, 44
Ireland in Pictures, 29
Irish Congress of Trade Unions (ICTU), 180
Irish Council for Social Housing (ICSH), 140
Irish Farmers' Association (IFA), 85
Irish Financial Services Regulatory Authority (IFSRA), 193
Irish Glass Bottle Company, 281, 284, 285
Irish Life Assurance, 75
   Abbey St development, *74, 75, 117*
Irish Nationwide, 118
Irish Rail, 248
Irish Times, 248, 284
Irishtown, 23, 151, 154, 160, 169–70, 184, 281
   Nature Reserve, 278
Isle of Dogs, 125, 137
Isle of Man, 114

Japan, 105, 116, 124
job placement programme, 194
Johnny Pedlars, 169
Johns, Ted, 136–7
Jury's Custom House Inn, 129, *130*, 297
Jury's Hotel Group plc, 129

Kavanagh, Mark, 109
Kealy, Professor Loughlin, 228
Kelly, Gerry, 150
Kelly, Patrick, 189
Kelly, Thomas, 29

Kelly family, 297
Keogh, Paul, 251
Kevany, Rose, 227–8
Kildare County Council, 68
Kilkenny, 70
Killarney, Co. Kerry, 183
KPMG Consultants, 126, 149, 179, 198, 199–200

La Touche House, 297
Labour Party, 94, 113, 114, 149, 163, 176, 296
Lambe, Seanie, 292
land ownership, 202, 288, 297
land rezoning, 155
Le Corbusier, 251
Leinster, duke of, 24
Liberties, 89, 131, 133, 166, 293
Liberty Hall, 75, 76, 243
Libeskind, Daniel, 273, 295
Liffey river, *48*, 133
   Bligh survey, 25, 28
   campshires, 154, 179, 288
   engineering works, 22
   land reclamation, 16–17, 22, 24, 33, 34, *45*
   northern development, 97, 111
   17th century channel, *17*
Limerick, 70, 298
Liverpool, 40, *43*, 89, 234–5
local elections, 2004, 155
Local Employment Charter, DDDA, 152, 177–8
Local Government, Department of, 49
Local Government (Planning and Development) Act 1963, 68, 70–1
Local Government (Planning and Development) Act 2000, 152, 164, 202, 206
local history projects, 49, 51
local population, 55, 67, 284–5
   'community capacity', 176–84
   effects of redevelopment, 164, 287, 292–3
   growth, 195–6, 202
   housing and employment, 152–3
   loss of heritage, 234–6
   non-national, 199
   Poolbeg Peninsula, 281, 282–4
   'socio-economic cleansing', 135–48
   and Spencer Dock, 248–9, 251–2, 255, 259, 262–3
   suspicions, 173–6
Locum Destination/At Large, 230
Loftus, Sean D. Dublin Bay-Rockall, 78
Lombard Court, 85
Lombard Street East, 85
Lombard Street West, 85

London, 42, 56, 107, 173, 292
   Canary Wharf, *98*, 99, 112, 116
   development and politics, 237, 253
   docklands development, 58, 89, 136–7, 212
   housing, 211
   pollution, 268
   rents, 118
   tax incentives, 125
   Tobacco Dock, 233–4
London Docklands Development Corporation, 57, 64, 99
Longboat Quay, 202, *204*
LUAS system, 248, 250, 293

M50 motorway, 214
M3 motorway, 214
MacAmhlaigh, Gus, 220, 221
McCann Fitzgerald, 271
McCarthy, Mark 215
McConnell, David, 252
McCreevy, Charlie, 225
McDaid, Jim, 240–1
McDonald, Frank, 247
MacGiolla, Tomás, 96
McGrath, Finian, 287
McGrath, Mairéad, 183, 184
McGuirk, Pauline, 70–1
McHugh, T.J., 85
Macken Street, 85, 250
McKenna, Patricia, 242
McManus, John, 284
McNamara, Bernard, 281, 284, 297
Magahy, Laura, 224
Maguire, Philip, 228–9
Mallin, Liavan, 199
Malone, Lorraine, 209
Maloney, Paul, 181
Malton prints, 23
MAN gasometer, 32
Manchester, 40
Mansergh, Martin, 73
marginalization, 37
Marine school, 23
Maritime Museum, Liverpool, 234–5
marketing campaigns, 57–8, 172–6
   heritage, 234–6
   housing, 196, 198
Marks & Spencer, 168
Master Project Agreement, 216, 218, 226
Mayor Square, 143, *144*, 205
Mayor Street, 22, 129, 168, *245*, 248, 249
   NCI campus, 189

Mayor Street Lower Urban Renewal Project, 140–3
Mayor Street Urban Renewal Brief, 174
Meath County Council, 68
Melbourne, 47, *47*
Melbourne Docklands Authority, 57
Memorial Road, 34, 97
Mespil Road, *77*
Millennium Commission, 239, 240
Millennium Dome, London, 237
Millennium Tower, *276*
Misery Hill, 273
Mitchell, Gay, 92
Mooney, Joe, 249
Moriarty Tribunal, 297
Mountjoy Square, *50*, *105*, 119–20
Mulcahy, Senator Noel, 73
Murphy, David, 216
Murphy, Kieran, 206
Murphy, Mahon, 221
Murray O'Laoire architects, 227
Museum of Dublin, 218, 228–31, 235
Museum of Modern Art, 218

National Action Plan on Poverty and Social Inclusion (NAPS Inclusion), 163
National Anti-Poverty Strategy (NAPS), 163
National College of Industrial Relations (NCIR), 187
National College of Ireland (NCI), 167, 183, 187–95, 205, 213, 248, 273, 297
National Conference Centre (NCC), 214, 237, 254, 256, 295
   contract finalized, 256, 257–8
   controversy, 240–2, 243–56
National Dance Company of Wales, 240
National Development Finance Agency (NDFA), 295
National Museums, Dublin, 230
National Spatial Strategy (NSS), 68–70, 265
National Sports Centre, 97, 111, 123, 132, 140
National Westminster Bank, 111
nationalization, 83, 89, 96, 135, 254, 287, 288, 296
NCB Stockbrokers, 111, 113
neo-liberalism, 148, 289
New Century House, 297
New International Division of Labour, 42
New Labour, 148, 163
New York, 42, 107, 121, 160, 173, 292
   rents, 118
   South Street Seaport, *62*, *63*, 102, 231
New York Cotton Exchange (NYCE), 120

Newfoundland Street, 22, *24*
Nixdorf, 116
Norfolk, Virginia, 63
North Block, IFSC, 116
North Bull Wall, 21, 28
North City Survey, 1919, 55
North Lotts, 22, 166, 209
North Port Dwellers Association, 248
North Strand, 52
North Wall, 184, 192, 262
　　extension, 31
North Wall Community Association, 109–10, 140–1, 143, 178, 210, 234–5, 250
North Wall Quay, *45*, 153, 154, 298; *see also* Spencer Dock; Stack A warehouse
North Wall Women's Centre, 182–3, 190
Northwestern Railway Hotel, 256, *257, 259*
Novell, 271

Ó Cofaigh, Tomas, 113
Ó hUiginn, Padraig, 113
O'Brien, Fergus, 93–4
O'Connell Bridge, 16, 93
O'Connell Street, 16, 240
O'Connor, Joyce, 187–8, 189–90, 193
O'Donoghue, Brendan, 149–50
O'Driscoll, Geraldine, 182
Odyssey Centre, Belfast, 61, 227
Office of Public Works (OPW), 240, 256, 295
Office of Science and Technology (OST), 227–8
O'Gorman, Brian, 200
oil crises, 79
Old Custom House, 20, *20*
Old Custom House Dock, *37*
Old Pigeon House Hotel, 280
Olympia and York, 116
Olympic Games, 56
Opera House Trust, Cardiff, 237, 239
Oracle Ireland, 125
Ordnance Survey, 1836, 22
O'Reilly, Joe, 273
O'Reilly, Marie, 248, 249
Oriel Street, 139
Ossory Road, 196
Ottawa, 268
Owens, Brian, 221

Paircéir, Seamus, 113
Parents in Education Initiative, 191
Park Hyatt Hotel Group, 273
Park West, 126
Parkman Consultants, 267
passenger ferries, 44

Pathways to Employment, 193–4
Pearse Square, 159, *161, 274*
Pearse Street, 52, 55, 85, 159–60, *161*, 168, 174, 187, 209
pedestrianization, 104
Percy Place, 126
Phoenix Park, 240
Pierse Construction Ltd, 205, 297
Pigeon House project, 225
Pigeon House Road, 160
Pittsburgh, 43, 58, 63, 173, 268
planning and development, 108
　　control of, 288
　　Custom House Docks, 97–101
　　framework, 67–71
　　and heritage, 214–15
　　and local residents, 96–7, 99–100
　　national strategy, 68–70
　　politics of, 66–108, 236–42
　　rezoning, 78–9
　　role of private sector, 289–91
　　Section 25 areas, 154–5
　　Spencer Dock controversy, 249–52
　　Urban Renewal Act, 94, 96–7
Planning Appeals Board, 236, 252
Planning Scheme, 1994, 111–12
Poddle Building, 224
Point Depot, 21, 99, 131, 132, 154, 248, 293
　　redevelopment, 295
Point Village Company, 295, 298
politics, and planning, 66–108, 236–42
Poolbeg Framework Plan, 282–4
Poolbeg Peninsula, 21, 151, 154–5, 209, 298
　　aerial view, *279*
　　difficulties, 213, 214, 265, 275–85
　　housing, *280*
　　land reclamation, 282
　　local population, 281, 282–4
Port Olimpico, Barcelona, 61
port tunnel, 28–9, 153
Post, An, 111, 121
Posts and Telegraphs, Department of, 34
Potsdamerplatz, Berlin, 237
Price Waterhouse, 113
PriceWaterhouseCoopers, 147, 256, 259
Princes Street South, 85
private sector investment, 35–6, 289–91, 297
privatization, 179, 212
　　of shoreline, 63, 148
　　urban space, 293–5
'privatopias,' 294
Progressive Democrats (PDs), 38, 283
project management, 93, 94, 96, 102

public-private partnerships, 57, 256
public sector intervention, 34–5

Quincy Market, Boston, 105
Quinlan, Derek, 297
Quinlan Partnership, 297
Quinn, Fergal, 180
Quinn, Lochlann, 297
Quinn, Ruairí, 113, 149–50, 179, 180, 182, 296

Rafferty, Mick, 110, 136
Rainbow Coalition, 163
Ranelagh, 187, 188, 189
residential development, 63
Reuters International News Agency, 195
Revenue Commissioners, 113
Revenue Dock, 28
rezoning, 78–9
Rhatigan, Brian, 121–2
Ringsend, 21, 23, 52, 151, 154, 174, 184, 281
    CCGS plant, 278
    flooding, 24
    housing, 85, 87, 160, 173
    industry, 54
Ringsend Park, 160
Riverside City, Limerick, 298
Roche, Kevin, 251
Rocque, map, 21
Rotterdam, 56, 212, 231
Rouse Company, 61
Royal Bank of Scotland, 111
Royal Canal, 37, 154
Royal Dublin Society (RDS), 240
Royal Hospital, Kilmainham, 218
Royal Institute of Architects of Ireland (RIAI), 92–3
RTÉ, 169

St Andrew's Resource Centre, 187
St James' Gate, 93–4
St Katharine's Dock, London, 58, 107
St Laurence O'Toole parish, 139
St Mark's Church, 160
St Pancras Housing Association, 139–40, 205
St Stephen's Green, 77
Saltaire, Bradford, 41
San Francisco, 132
Sandymount and Merrion Residents Association, 72
Saolscoil, 186–7, 188, 189
Sassen, S., 292
'scanscapes,' 294
Schools Job Placement Programme, 177–8

Schwarz, Martha, 273, 295
Scientazia, 220
Scott Tallon Walker, 73
Sean McDermott Street Residents' Association, 178
Sean Moore Park, 160, 281
Seanad Éireann, 93–4, 96, 101, 150, 163, 180
    Wood Quay, 215, 235
Section 25 areas, 154–5, 243, 246, 256, 259
    Poolbeg Peninsula, 281
security, 174, 201–2
Seville Place, 91, 192–3, 209
sewage project, 21
Shanagher, Martin, 227
Shannon Foynes Port Company, 298
Sheridan, Nicky, 125
Sheriff Street, 22, 34, 49, 96–7, 110, 162, 168, 170, 209
    aerial view, *142*
    housing, *51*, 52, 53–4, 111, 198–9, 205
    playground, 141, *142*
    redevelopment, 292
    security, 174
    'socio-economic cleansing,' 135–48
    sorting office, 121
    unemployment, 55
    'wall,' *146*
Sheriff Street Bridge, 256, 259
Sheriff Street Youth Club, 145, 178
Sinn Féin/The Workers' Party, 55
Sir John Rogerson's Quay, 23, 24, 31, 159, 269, 274
    gasometer, 32
Sites and Monuments Record, 159
Skelly, Liam, 94
slob lands, 22
Smith, Michael, 251
Smithfield, 133, 189, 293
social change, 165–72
social inclusion, 37, 63–4
Social Inclusion Agenda, 163
Social Inclusion Units, 148
social polarization, 63–5, 145, 162, 164, 173–6, 210–11, 292–3
Sonas consortium, 240
South Bank Quay, 274, *278*
South Dublin County Council, 68, 283
South Lotts, 23, 32
South Street Seaport, New York, *62*, 63, 102, 231
South Wharf plc, 281
'spatial fix,' 65
Spencer Dock, 97, 131, 153, 155, 202, 284
    aerial view, *260*

and DDDA, 290
development controversy, 243–56
difficulties, 189, 213, 214, 236, 237, 281
environs, *238*
housing, 140, 200, 209–11
and local community, 176, 177–8, 248–9, 251–2, 259, 262–3
marketing, *261*
NCC proposal, 240–2, 295
oral hearing, 147
politicized, 252–6
present condition, 256–63, 291
revised development, *258*
Spencer Dock Development Company, 243–56, 254, 288
Sport and Tourism, Department of, 240
Stack A, 76, 106–7, 147, 213–14, 254, 273, 284
chq, 231–5
control of, 294
development difficulties, 215, 216–36
heritage dispute, 215, 216–36
influence of DDDA, 288–9
Museum of Dublin, 228–31
Science Centre, 218–28
Station Square, Pittsburgh, 63
Strategic Planning Guidelines for the Greater Dublin Area, 68
Sumitomo Bank, 116
Summerhill, *50*, 91
Sun Microsystems, 125, 126
'super-blocks,' 159
super-trucks, 29
Swedish Food Company, 168
Swift, Jonathan, 221, 236
Sydney, 58, *59*, 61, 132, 268
Sydney Opera House, 295

Taisce, An, 71–2, 89, 251
Talbot Memorial Bridge, *74*
Tallaght, 79, 208
Taoiseach, Department of the, 193
Tara, 214
tax incentives, 93, 149, 151, 166, 251
benefits and costs, 126–8
and IFSC, 113–14, 123–4
Taylor, Peter, 286
Techniquest Science Museum, Cardiff, 102, *103*, 219–20, 227
Technopole, TCD, 280
Temple Bar, 243
Temple Bar Properties (TBP), 224, 291
Tempozan, Japan, 105
Thatcher, Margaret, 64, 148

Third-level Scholarship, 195
Thompson, Benjamin, 104, 129, 152, 295
Thorncastle Street, 52, *52*, 53, 160, 208–9
Threshold, 206
Ticknell, Rob, 262
Tobacco Dock, Wapping, 233–4
Tokyo, 37, 42, 118
Toronto, 63, 179
Tourism, Sport and Recreation, Department of, 240–1
Town and Country Planning Act, UK, 1947, 67, 68
Townsend Street, 85, *86*
traffic problems, 153–4
Transport, Department of, 241
Treasury Holdings, 240, 242, 248, 249, 251–2, 259, 274
Spencer Dock, 243–56, 247, 256
Tribunals of Inquiry, 66, 297
Trinity Access Programme (TAP), 187
Trinity College, Dublin (TCD), 104, 180, 243
outreach, 187
Technopole, 280

U2 Tower, 273
UCD College of Business and Law, 227
Ulster Bank, 118
unemployment, 49, 51, 55, 64, 80, 82, 83
docklands, 170–1
United Air, 125
United Nations, 56, 68
United States of America (USA), 61, 102, 116, 124, 131
economic decline, 42–3
environment, 263
marketing, 173
new social agenda, 163
planning, 70
Urban Development Corporations, 93, 289
urban revitalization, 148
University College Dublin, 148, 228
University of Loughborough, 286
Urban Development Areas Bill, 1982, 87, 88–92, 91–2, 93
Urban Development Corporations (UDCs), 93, 289
urban regeneration, 22, 39–40, 58, 101, 286, 298
commissions, 81
and 'community capacity,' 176–84
counter-urbanization, 265
and local population, 164, 173–6
politics of, 252–6
privatization, 293–5

urban regeneration (*contd*)
　problem areas, 263–7
　social polarization, 63–4
　transformation, 40–4
　use of waterfronts, 56–65
Urban Renewal Act 1986, 92–7, 108, 109, 135
　boundary changes, 107, 111
Urban Renewal Development Brief, 140

V&A Waterfront, Cape Town, 63

W5 Centre, Belfast, 227
Wales Millennium Centre, 239–40
Walsh, Brendan, 148–9
warehouse conversions, 58
Warren, Kevin, 297
Waste to Energy Facility, 282
waterfront regeneration, 92, 101, 298
　Dublin as model, 211–12
　economic decline, 44–9
　marketing, 173
　privatization, 179
　problem areas, 263–7
　redevelopment of, 56–65
　social polarization, 63–5, 145

Waterside, Norfolk, Virginia, 63
Waterways Ireland, 180
Wellington Quay, 20
Welsh National Assembly Building, 102
Welsh National Opera, 237, 240
Welsh Office, 237, 239
Westland Row, 151, 154, 159, 168
Wicklow County Council, 68
Wide Streets Commission, 40
Wiggins Teape factory, *156*
Winter Garden, Battery Park City, 105
Wood Quay, 17, 94, 215, 235
Workers Party, 96
World Bank, 56
World Cup, 56
World Health Organization, 56
Wright Plan, 49, 68

Yale University, 191
York Road, 132
Young Mothers' Self Development Programme, 182–3

Zelda properties, 208–9
Zoe Developments, 131–3